27 Advances in Polymer Science
Fortschritte der Hochpolymeren-Forschung

Edited by H.-J. Cantow, Freiburg i. Br. · G. Dall'Asta, Cesano Maderno
K. Dušek, Prague · J. D. Ferry, Madison · H. Fujita, Osaka
M. Gordon, Colchester · W. Kern, Mainz · G. Natta, Milano
S. Okamura, Kyoto · C. G. Overberger, Ann Arbor · T. Saegusa, Kyoto
G. V. Schulz, Mainz · W. P. Slichter, Murray Hill · J. K. Stille, Fort Collins

With 97 Figures

Springer-Verlag
Berlin Heidelberg New York 1978

Editors

Prof. Dr. HANS-JOACHIM CANTOW, Institut für Makromolekulare Chemie der Universität, Stefan-Meier-Str. 31, 7800 Freiburg i. Br., BRD

Prof. Dr. GINO DALL'ASTA, SNIA VISCOSA – Centro Sperimentale, Cesano Maderno (MI), Italia

Prof. Dr. KAREL DUŠEK, Institute of Macromolecular Chemistry, Czechoslovak Academy of Sciences, 162 06 Prague 616, ČSSR

Prof. Dr. JOHN D. FERRY, Department of Chemistry, The University of Wisconsin, Madison 6, Wisconsin 53706, U.S.A.

Prof. Dr. HIROSHI FUJITA, Osaka University, Department of Polymer Science, Toyonaka, Osaka, Japan

Prof. Dr. MANFRED GORDON, University of Essex, Department of Chemistry, Wivenhoe Park, Colchester C04 3 SQ, England

Prof. Dr. WERNER KERN, Institut für Organische Chemie der Universität, 6500 Mainz, BRD

Prof. Dr. GIULIO NATTA, Istituto di Chimica Industriale del Politecnico, Milano, Italia

Prof. Dr. SEIZO OKAMURA, Department of Polymer Chemistry, Kyoto University, Kyoto, Japan

Prof. Dr. CHARLES G. OVERBERGER, The University of Michigan, Department of Chemistry, Ann Arbor, Michigan 48 104, U.S.A.

Prof. TAKEO SAEGUSA, Kyoto University, Department of Synthetic Chemistry, Faculty of Engineering, Kyoto, Japan

Prof. Dr. GÜNTER VICTOR SCHULZ, Institut für Physikalische Chemie der Universität, 6500 Mainz, BRD

Dr. WILLIAM P. SLICHTER, Bell Telephone Laboratories Incorporated, Chemical Physics Research Department, Murray Hill, New Jersey 07 971, U.S.A.

Prof. Dr. JOHN K. STILLE, Colorado State University, Department of Chemistry, Fort Collins, CO 805 23, U.S.A.

ISBN 3-540-08829-6 Springer-Verlag Berlin Heidelberg New York
ISBN 0-387-08829-6 Springer-Verlag New York Heidelberg Berlin

Library of Congress Catalog Card Number 61-642

This work is subject to copyright. All rights are reserved, whether the whole or part of the material is concerned, specifically those of translation, reprinting, re-use of illustrations, broadcasting, reproduction by photocopying, machine or similar means, and storage in data banks. Under § 54 of the German Copyright Law where copies are made for other than private use, a fee is payable to the publisher, the amount to the fee to be determined by agreement with the publisher.

© by Springer-Verlag Berlin Heidelberg 1978
Printed in Germany

The use of general descrive names, trademarks, etc. in this publication, even if the former are not especially identified, is not to be taken as a sign that such names, as understood by the Trade Marks and Merchandise Marks Act, may accordingly be used freely by anyone.
Typesetting and printing: Schwetzinger Verlagsdruckerei. Bookbinding: Brühlsche Universitätsdruckerei, Lahn-Gießen.
2152/3140 – 543210

Preface

Polymer science and engineering as a discipline is some fifty years old. This brief lifetime has seen the development of synthetic elastomers that equal or exceed nature's product, Hevea rubber, in abrasion resistance, tensile strength, high temperature performance, and degradation resistance; the development of a molecular theory of rubber elasticity, truly a triumpf of statistical mechanics; the development of synthetic fibers that now clothe a significant fraction of the world's population; the emergence of plastics as structural or protective elements for the sheltering of man; the use of polymeric films and materials for artificial hearts, kidneys and blood dialysis; the synthesis of stereospecific polymers which come close to the duplication of nature in chemical modeling; and countless other areas where low density, optical clarity, dielectric activity (or the lack of it), corrosion resistance, biological inertness, ease of fabrication, or other specific properties dictate the use of high polymers.

Whereas polymer organic chemistry represented the major academic endeavor during the early years of macromolecular science, the last twenty years have indicated a trend toward the emphasis of polymer physics and physical chemistry. The last several years give clear indication that a major re-emphasis is about to occur once again in that the field of polymer engineering is beginning to emerge. Industry has a clear need for engineers and scientists versed in the engineering sciences but with expanded knowledge of the properties and processing of polymers. In particular, problems associated with the failure of polymers, such as the engineering properties of fracture, creep resistance, impact strength, fatigue and solvent stress cracking and crazing are numerous and difficult.

The statistical structure of polymeric glasses and the broad spectrum of order-disorder and morphology in "crystalline" polymers have yet to be quantified to the degree to which defects such as vacancies and dislocations have been quantified for metals. This, together with the strong dependence of polymer solid properties on the melt rheology and history, as compared to the weak dependence of metal properties on melt history, makes the relationships between failure properties and "structure" of polymeric solids difficult and often elusive. As is well known, small but significant changes in orientation of the solid resulting from changes in melt flow field (*e.g.* by changes in die design) can lead to greatly improved or reduced tensile strength or impact strength.

What is clear is that specification of the structure is far more complex than a delineation of chemical composition, tacticity, molecular weight and so forth. Recent studies on glassy polymers have shown that thermal history is a primary variable for these non-equilibrium materials. However, the extent to which gross or subtle changes in morphology as induced by the stress, strain, temperature and flow histories of the solid and melt precursor affect the ultimate properties of the solid remains to be delineated. As polymers are used with increasing regularity in structural engineering components, it will become of major importance both to control their properties through a more thorough understanding of the relevant structural parameters of the final solid object and to design (in the engineering context) with meaningful mechanical properties data which reflect the strongly time, temperature and stress state and level dependent properties which polymers exhibit.

The three articles which appear in this volume represent distinct, but complementary aspects of the general theme of failure in polymers. Professor Andrews has summarized research on the molecular failure mechanism itself as reflected in radical formation which occurs during chain scission. As he points out there is considerable difficulty in correlating directly the rate of radical formation with the applied stress or strain levels and time histories. In part this is due to experimental difficulties associated with performing stress-strain experiments in the spectrometer cavity. A number of the studies reported were, unfortunately, not very specific as to the sample stress or strain and loading history. Furthermore, catastrophic failure as embodied in the fast propagation of a crack is most evasive in that the localization of the radicals produced does not lead to significant sensitivity in the spectrometer cavity. Thus, the technique has proved to be most useful for ubiquitous production of radicals throughout the sample.

From a conceptual viewpoint the primary theoretical problem yet to be solved is the stress transfer mechanism in polymer solids. As noted earlier, polymers have statistical structures when in the glassy state and a rather broad spectrum of order-disorder when in the crystalline state. Detailed analysis of stress transfer through a glassy structure requires comprehensive analysis of chain conformation in the (nonequilibrium) glass which in turn requires an understanding of both the intramolecular and intermolecular energetics.

Crystalline polymers appear to be the most studied by ESR techniques. The model which seems to emerge from these results is, in fact, a variant of a model proposed over twenty years ago by Cumberbirch and associates (Shirley Institute Memoirs) to explain the tenacity of wet rayon monofilaments. Briefly, Cumberbirch, *et al.* propose a fringe-micelle structure in which the fringe regions, swollen by water, are assumed to obey rubber elasticity theory. These fringe regions are, of course, the more accessible (to water), more disordered, regions of the semicrystalline structure.

A statistical distribution of connector chain lengths, which depends on both the micelle spacing and the distribution of chain lengths (degree of polymerization), connects the micelles. As stress is applied to the sample the average spacing between micelles changes and results in nonuniform strain among the connector chains. Cumberbirch then invokes a taut chain failure criterion and calculates the average strain at which the unbroken chains can no longer accept the extra stress imposed on them by the rupture of the next taut chain. The failure process described by Cumberbirch is in essence the model which seems to be in reasonable agreement with the ESR studies of failure in crystalline polymers.

While the fringe-micelle model for crystalline polymers has not been fashionable for some time, it may have some utility in modeling stress transfer and failure mechanisms. In any event, a fringe micelle model is a primitive form of more general composites models which attempt to model the behavior of crystalline polymers using the same techniques as for filled systems or fiber reinforced plastics. The ESR studies may serve to provide valuable insight into the validity of such models for crystalline polymers particularly in regard to the manner in which stress is shared or distributed in the more disordered regions.

Applications of linear elastic fracture mechanics (primarily) to the brittle fracture of solid polymers is discussed by Professor Williams. For those not versed in the theory of fracture mechanics, this paper should serve as an excellent introduction to the subject. The basic theory is developed and several variants are then introduced to deal with weak time dependence in solid polymers. Previously unpublished calculations on failure times and craze growth are presented. Within the framework of brittle fracture mechanics and testing this paper provides for a systematic approach to the failure of engineering plastics.

Several cautions are, however, in order. Polymers are notorious for their time dependent behavior. Slow but persistent relaxation processes can result in glass transition type behavior (under stress) at temperatures well below the commonly quoted dilatometric or DTA glass transition temperature. Under such a condition the polymer is ductile, not brittle. Thus, the question of a brittle-ductile transition arises, a subject which this writer has discussed on occasion. It is then necessary to compare the propensity of a sample to fail by brittle crack propagation versus its tendency to fail (in service) by excessive creep. The use of linear elastic fracture mechanics addresses the first failure mode and not the second. If the brittle-ductile transition is kinetic in origin then at some stress a time always exists at which large strains will develop, provided that brittle failure does not intervene.

An additional complication for glassy polymers is their spontaneous aging for many years following vitrification. Linear elastic fracture mechanics can only treat the crack propagation parameters that currently prevail in the test specimens.

For the reasons cited, it is prudent to evaluate plastics for long term stress-supporting applications using linear elastic fracture mechanics in conjunction with other rheological and thermophysical data, particularly regarding long time behavior, aging phenomena, and failure modes.

Failure in multiphase polymers and polymer composites (non-fibrous) is reviewed by Professor Bucknall. Several examples are presented in which the effect of adding a dispersed second phase to a polymer can be either beneficial or deleterious to stress, strain, or work to break. It is shown that two basic modes of local plastic deformation may be operative, namely crazing and shear band formation. By studying the sample dilatation versus strain in uniaxial tension creep tests, Bucknall is able to determine the operative mechanism in each system. Fracture mechanics is used to evaluate the toughness parameters of the various systems.

It is noted that attempts to apply composites theory to the materials investigated have not been entirely successful. While upper and lower bounds on, *e. g.,* moduli can be established there is little quantitative prediction of the impact strength or fracture toughness parameters of the composites. Hence, the systems cannot be considered as optimized, for example, with regard to impact strength versus particle size, shape, or distribution or matrix-particle adhesion. The complexity is, of course, due to the statistical structure of the dispersed phase and the resultant uncertainties in the calculations of *local* stress fields, which in turn imply uncertainty in the local mode of yielding or rate of yielding.

Conceptually, the problems associated with the optimization of specific mechanical properties by variations of structure and morphology are the same in rubber-filled systems, glass-bead filled systems and semicrystalline polymers. When the fracture properties are singled out, our understanding of the relationships between macroscopic failure and local failure is hampered by the limited knowledge of stress transfer in statistically nonhomogeneous structures. The increased use of composites theory and micromechanics to address these problems would appear to be appropriate.

Professors Andrews, Williams, and Bucknall have summarized the current status of the molecular, phenomenological, and materials aspects of failure in polymers, respectively. Any future developments in the linkage among these three approaches will, of necessity, serve to enhance each of them with the knowledge of the others.

Materials Engineering Department
Rensselaer Polytechnic Institute
Troy, New York 12181, U.S.A.

S. S. Sternstein

Contents

Molecular Fracture in Polymers
   E. H. ANDREWS and P. E. REED   1

Applications of Linear Fracture Mechanics
   J. G. WILLIAMS   67

Fracture and Failure of Multiphase Polymers
and Polymer Composites
   C. B. BUCKNALL   121

Author Index Volumes 1–27   149

Molecular Fracture in Polymers

Edgar H. Andrews and Peter E. Reed

Department of Materials, Queen Mary College, London E1 4NS, Great Britain

Table of Contents

1. Approaches to Fracture 3
   1.1. Continuum Mechanics 3
   1.2. The Nature of the Surface Energy 6
   1.3. The Kinetic Theory of Fracture 8
   1.4. Mathematical Formulation of the Kinetic Theory 10

2. Experimental Study of Molecular Fracture 13
   2.1. Infra-red Spectroscopy 13
   2.2. Low Angle X-ray Studies 16
   2.3. Electron Spin Resonance — Basic Concepts 17
   2.4. Paramagnetic Centre Formation in Polymers 21
   2.5. Limitations on the Application of ESR to Polymer Studies . . . 21
   2.6. Experimental Procedures 24

3. Mechanical and Physical Processes 27
   3.1. Factors Controlling Molecular Fracture 27
   3.2. The Importance of Molecular Anchorage 28
   3.3. ESR Signals from Uncross-linked and Amorphous Polymers . . . 28
   3.4. Crystalline Fibre Structures 29
   3.5. Glassy Polymers and the Role of Cross-links 31
   3.6. The Role of Chemical Structure 33
   3.7. The Roles of Stress and Strain 33
   3.8. The Effects of Pre-Strain and Plastic Strain 36
   3.9. The Effects of Strain Rate 40
   3.10. The Effects of Temperature 41
   3.11. Crazes and Microvoids 43
   3.12. Synopsis of Mechanical and Physical Processes 46

4. Radical Formation, Identification and Decay 48
   4.1. Identification of Radical Species 48
   4.2. Primary and Secondary Radical Species 53

4.3.	Reactions with Oxygen	57
4.4.	Radical Decay	61

5. Conclusion 63

6. References 64

1. Approaches to Fracture

The subject of fracture is of both practical and scientific interest. The propensity of a material to fracture sets limits both to engineering design and to the service life of engineering components and structures. Questions of safety, as well as convenience, are deeply involved.

At the same time the phenomenon of fracture reflects, in one way or another, the ultimate limit of deformation in a solid. It thus involves fundamental physical properties of the material such as its inter-atomic bonding, its surface energy and its crystal structure. It also involves crystallographic processes such as slip, stress induced phase transformations and twinning, whilst in molecular solids such as polymers other processes such as molecular relaxation behaviour may predominate. Fracture is clearly of great scientific interest and has attracted the attention of chemists and solid state physicists as well as engineers.

It is not surprising therefore that the subject has, historically, been approached from a variety of different viewpoints. In this introductory section we review briefly these various 'approaches to fracture' and point out the methods by which they can be co-ordinated and, where necessary, reconciled with one another.

1.1. Continuum Mechanics

This approach is the most useful for engineering purposes since it expresses fracture events in terms of equations containing measurable parameters such as stress, strain and linear dimensions. It treats a body as a mechanical continuum rather than an assembly of atoms or molecules. However, our discussion can begin with the atomic assembly as the following argument will show. If a solid is subjected to a uniform tensile stress, its interatomic bonds will deform until the forces of atomic cohesion balance the applied forces. Interatomic potential energies have the form shown in Fig. 1 and consequently the interatomic force, which is the differential of energy with respect to linear separation, must pass through a maximum value at the point of inflection, P in Fig. 1.

This maximum load-bearing capacity of the atomic bond can be expressed[1] as a breaking stress for the bond, thus,

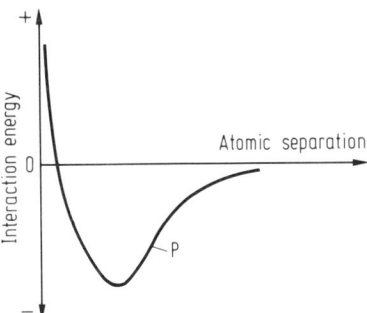

Fig. 1. Interatomic potential energy (schematic)

$$\sigma_m \simeq E\epsilon_m \tag{1}$$

where E is the Young's modulus of the solid, and ϵ_m is the strain at which σ_m is achieved. Since most laws of force between atoms joined by primary bonds give an ϵ_m of 0.1 to 0.2, the bond fracture stress can be written

$$\sigma_m = \alpha E \tag{2}$$

$$0.1 < \alpha < 0.2$$

The simplest continuum approach to fracture is to ascribe to the continuous solid the strength of its individual atomic bonds. Thus,

$$\sigma_f = \sigma_m = \alpha E \tag{3}$$

where σ_f is the macroscopic fracture stress.

It is well known that this "theoretical strength" is seldom, if ever, achieved by solids, although freshly drawn glass fibres and certain whisker crystals do appear to exhibit tensile strengths approaching the theoretical limit.

Bulk solids, generally, exhibit α values in Eq. (3) between 10^{-2} and 10^{-4}, that is one to three orders of magnitude smaller than expected, and this is ascribable to flaws, cracks and imperfections in the body which concentrate stress[2].

The idea that the strength of bulk solids is controlled by flaws was advanced by Griffith[3] in 1921 and has led to the development of a much more sophisticated continuum approach to fracture, known as fracture mechanics. Fracture mechanics is concerned always with the conditions for the propagation of an existing crack, and it is important to bear this in mind when comparing different theories of fracture. Griffith's ideas are well known and do not need to be elaborated here. There are some aspects of his theory which are relevant to the present discussion, however. Griffith's equation for the fracture stress of an elastic material is (for plane stress),

$$\sigma_f = \sqrt{\frac{2ES}{\pi c}} \tag{4}$$

where S is the surface energy of the solid and c the length of the largest pre-existent crack. For this equation to give the correct limit as $c \rightarrow 0$, the term c should be replaced by $(c + d)$ where d is the interatomic spacing. Furthermore, the surface energy S needs to be carefully defined. Strictly, it is half the energy required to fracture unit area of inter-atomic bonds across the fracture plane, this plane being created by propagation of the pre-existent flaw.

The important point about Griffith's theory is that is *does* contain the bond fracture energy, or bond strength, explicitly in the term S. This distinguishes the theory from all subsequent theories of fracture mechanics with the exception of the "generalized theory" recently proposed by Andrews[4]. Thus although Griffith's is a continuum theory it does relate directly to atomistic parameters.

Unfortunately the theory is derived for purely elastic solids and cannot handle the inelastic deformations (and thus energy losses) which normally predominate in the vicinity of a crack tip nor the time and temperature dependence commonly found in fracture. As a result Griffith's equation predicts fracture stresses much lower than are normally observed, even in brittle materials.

Two methods have been adopted to overcome this problem. In the first, due to Orowan[5] and others[6], the term S is generalized to include dissipative energy contributions which are thus, somewhat arbitrarily, assigned to the new crack surfaces although they actually occur over a volume distribution around the crack tip and even throughout the stressed body. In this paper we shall use the expression $\mathscr{T}$ to represent this modified "surface energy" and we shall call it the "surface work".

The difficulty here is that $\mathscr{T}$ is no longer related in a discernable manner to the physical properties of the solid and must be treated as an empirical quantity which, hopefully, is constant for a given material. As it happens, $\mathscr{T}$ is found to vary with such things as rate, temperature and sheet thickness.

The second approach, due to Irwin[7], is to characterise the stress field surrounding a crack in a stressed body by a stress-field parameter K (the "stress intensity factor"). Fracture is then supposed to occur when K achieves a critical value K_c. Although, like Griffith's equation, this formulation of fracture mechanics is based on the assumptions of linear elasticity, it is found to work quite effectively provided that inelastic deformations are limited to a small zone around the crack tip. Like $\mathscr{T}$, however, the critical parameter K_c remains an empirical quantity; it cannot be predicted or related explicitly to the physical properties of the solid. Like $\mathscr{T}$, K_c is time and temperature dependent.

Thus, whilst the science of fracture mechanics has flourished as a means of defining fracture properties for engineering purposes, it remains basically empirical; the critical parameters (like $\mathscr{T}, K_c, J_c, \delta_c$) cannot be predicted from, or related to, the physical properties of the solid in question.

This situation has been resolved, at least in part, by a recent generalized theory of fracture mechanics[4] which gives,

$$\mathscr{T} = S \Phi (\dot{c}, T, \epsilon_0) \tag{5}$$

Here $\mathscr{T}$ is again the surface work, S is the surface energy as previously defined and Φ is the "loss function" dependent on crack speed, temperature and the strain, ϵ_0, applied to the specimen. The theory gives Φ explicitly in terms of the energy density distribution in the specimen and the plastic or visco-elastic hysteresis of the material.

In principle, then, the surface work $\mathscr{T}$, which determines the fracture stress of the body can be calculated from the physical properties of the material. In practice this is not easy, since the energy density distribution can only be calculated exactly for linear elastic solids, for which $\Phi \to 1$ and Eq. (5) reverts to the Griffith theory. However, Eq. (5) has been found correct for a series of highly extensible materials in which the energy density distribution was measured experimentally[8].

The generalized theory therefore restores the explicit link between a continuum mechanics approach to fracture, which is of such great value in engineering design and practice, and the atomistic view which concerns us most in this review. This link has been lost since Griffith's theory was found to be inadequate for most real materials, but is now reinstated by means of the same parameter employed by Griffith, namely the surface energy S.

1.2. The Nature of the Surface Energy

We now examine more closely the significance of the term S. As previously noted, this is not the quantity normally referred to as surface energy and which controls *e.g.* the contact angles of liquids on solid surface. The latter (we will denote it γ_s, or γ_{sv} if the solid is in contact with a vapour phase as is usually the case) only corresponds to S in rather unusual circumstances, such as ultra-clean metal surfaces in vacuum. For polymeric solids, γ_s reflects only the weak secondary bonding that exists between molecules, whereas S (the energy to break unit area of bonds across a fracture plane) normally contains major contributions from primary bond rupture.

For metals and ionic solids, in which atoms interact only by omni-directional primary bonds, it is clear that S will be the fracture energy of such bonds normalised to unit area. For co-valently bonded solids, like diamond, the secondary bonding energies are negligible with respect to the primary bond strengths so that S will be given directly by the latter — again normalised to unit area.

Molecular solids like polymers present greater problems since the creation of surfaces may involve the severence of primary (intra-molecular) or secondary (intermolecular) bonds, or, more likely, both simultaneously. In thermoplastics it is possible to envisage molecular "pull-out" in which no molecules are broken but are simply separated from one another against the frictional secondary bonding forces. In network polymers, of course, surfaces can only be created by breaking primary bonds, but these may be relatively widely spaced. Crystalline thermoplastics and indeed amorphous polymers with very long molecules may behave (in this respect) more like network polymers because crystals or entanglements act as effective cross-links.

A theory due to Lake and Thomas[9] appears to provide a satisfactory account of the origin of the parameter S in cross-linked systems. (In the literature the symbols $T_0 \equiv 2S$ and $\mathcal{T}_0 \equiv S$, are used, indicating minimum or threshold "tearing energy" referred respectively to unit area of fracture plane *i.e.* two surfaces, and unit area of fracture surface).

Lake and Thomas supposed that no primary bond in a cross-linked network can fracture unless all the bonds in that particular network chain (*i.e.* between adjacent cross-links) are stressed to breaking point. Thus, if there are $\bar{n}$ inter-atomic bonds between cross-links (on average) the minimum energy to cause *one* bond to fracture is not the dissociation energy of one bond but n times that energy. Their prediction takes the form

$$S = \bar{L} N \bar{n} \, U/2 \tag{6}$$

where $\bar{L}$ is the mean displacement length between cross-links (*i.e.* the distance between cross-links in the unstrained state), N is the number of network chains per unit volume and U is the dissociation energy of a single bond.

The theory predicts S values in the range $10-20$ Jm^{-2} for normal cross-linked elastomers, depending of course on cross-link density. This compares favourably with minimum tearing energies measured experimentally by fatigue-limit observations[10] and other special techniques and which lie in the range $20-50$ Jm^{-2}. A modified form of the theory was used by King[11] to calculate S for epoxy resin networks and again agreement with experiment was obtained within a factor of two. These experimental measurements of S are possible for cross-linked elastomer networks only because, at low rates of crack propagation and elevated temperatures, visco-elastic losses can be reduced to zero, giving $\Phi \to 1$.

A different approach to the determination of S was adopted by Andrews and Fukahori[8] who treated S as the unknown in Eq. (5) and found values of 65 Jm^{-2} for SBR and EPDM rubbers and 100 to 200 Jm^{-2} for polyethylene and plasticized PVC. The latter represent the first determinations of S for non cross-linked polymers.

The values cited are, of course, of the order of 10^3 times γ_{sv} for polymeric solids, emphasizing the difference between S and γ. They are also up to 100 times the surface energy values to be expected from "high surface energy" solids such as clean metals and oxides (< 0.5 Jm^{-2}). It is clearly necessary to invoke the mechanism of Lake and Thomas to explain these high S values in polymeric materials.

Under some circumstances it is possible to observe very low S values for polymer solids, namely when the energy to fracture chemical bonds is provided by chemical reaction. An excellent example of this is the ozone cracking of unsaturated hydrocarbon elastomers where crack propagation occurs at threshold values of $\mathcal{T}$ as low as 0.05 Jm^{-2} [12]. Under such low stresses the loss function of Eq. (5) is effectively unity and $\mathcal{T} \sim S$. Then S is probably nothing more than the surface energy of the (liquid) degraded rubber at the tip of the crack.

Low values have also been recorded[13] in the solvent stress-crazing of glassy plastics, where in appropriate circumstances $S \sim 0.1$ Jm^{-2}. In this case, however, the surface energy in question is that of voids and cavities in the craze.

To summarize, therefore, we conclude that the fracture resistance of solids as reflected in the parameter $\mathcal{T}$, is controlled by

(i) the mechanical energy losses which determine Φ and
(ii) the energy required to sever interatomic bonds across the fracture plane.

In cross-linked polymers, crystalline polymers and high molecular weight glasses, the bond fracture energy per area, S, is dominated by the strength of co-valent intramolecular bonds. The only exception is where the energy to rupture these bonds is supplied by extraneous chemical means rather than by mechanical stress. Although a higher concentration of cross-links, tie points or entanglements increases the number of co-valent bonds to be broken in unit area, this effect is more than offset by the Lake-Thomas effect which requires an entire network chain to be energised to break for each bond that actually ruptures. Thus, paradoxically, the fewer the cross-links, the higher does S become until, of course, the contribution to S of secondary bonds begins to predominate when S will fall again.

Further studies of the significance of S in polymer fracture will certainly be carried out in future research. The main purpose of the forgoing discussion, however, is to emphasize that molecular fracture is not a scientific curiosity but plays a major role in determining the resistance of polymer solids to failure. Hopefully, the use of Eq. (5) will make it possible to quantify this role even in the field of engineering fracture mechanics.

1.3. The Kinetic Theory of Fracture

Developed chiefly in Russia, the kinetic theory of fracture at first appears to represent an entirely different account of fracture phenomena to that discussed in Section 1.2. In fact some Russian authors have claimed that the kinetic theory contradicts the Griffith theory of fracture[14]. As we shall see, however, this is not the case.

In the kinetic theory, attention is focused on the event of bond fracture and the latter is described in terms of chemical rate theory. The mechanical stress on the bond (normally the result of an applied load) modifies the free energy barrier that must be crossed if the bond is to change from an unbroken to a broken state. Figure 2 illustrates the situation.

In Fig. 2a, which represents, the unstressed state, the transition,

bond intact (A) ⟶ bond broken (B)

is seen to require both an activation energy G^* and a net increase in free energy ΔG. Because ΔG is positive no net reaction A → B can occur, but if it *could*, and if all

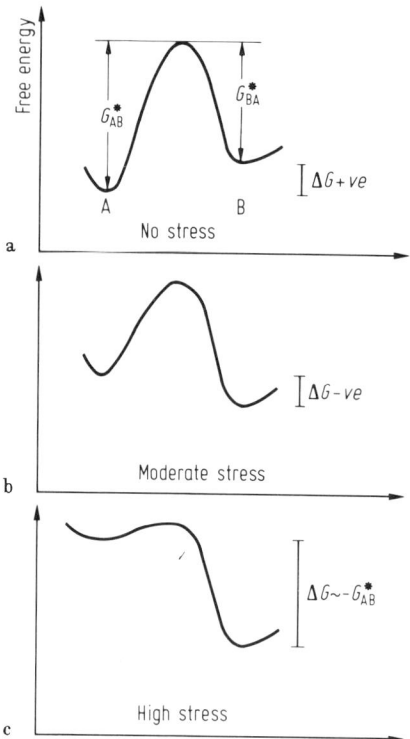

Fig. 2. The reaction (bond intact, state A) → (bond broken, state B) and the effect of tensile stress

broken bonds lay in the same plane, the free energy increase ΔG would represent the surface energy of that plane, (when normalised to the appropriate area). The magnitude of ΔG may be affected by reaction of the broken bond with environmental species such as oxygen.

The surface energy represented by ΔG is of the nature of γ_s referred to earlier. In contrast the fracture surface energy S is related to the activation energy G_{AB}^*, since this is the energy that must be supplied to break the bond in the first instance.

Figure 2b shows the situation when the bond is subject to a moderate tensile stress. Now ΔG is just negative and a net "forward" reaction A → B is to be expected. If we consider this to be occurring at the tip of a propagating crack, the effective population of states B will never rise to a significant level in the "process zone" because these states are removed from the reacting system by the translation of this zone (i.e. the crack tip) through the solid.

The point to notice is that a threshold level of stored energy in the bond is necessary before $\Delta G \to 0$ and thus before the "forward" reaction can proceed. This stored energy is, of course, equal to ΔG and thus (when normalised) to the surface energy of the material. Thus we see that the kinetic model of fracture requires a minimum energy supply equal to the surface energy, which is the fracture criterion employed by Griffith. No contradiction therefore exists between the kinetic and Griffith theories.

The stress conditions which give $\Delta G < 0$, and which therefore permit crack growth to occur by bond fracture at the tip, may be divided into two regimes. The first (low stress) regime occurs when $-\Delta G \ll G^*$, as is the case for Fig. 2b. Here bond rupture, and hence crack growth, can only take place as a result of large thermo-fluctuations and the rate of growth will be small and highly temperature dependent. This will generally obtain at low applied stresses and also at very small crack length, since the shorter the crack, the lower is the concentration of stress at its tip. This enables us to identify events occurring in this regime, broadly though not exclusively, as "initiation" events.

In contrast, as Fig. 2c shows, the stress on the bond may be so high that $-\Delta G \simeq G^*$, and only very small thermo-fluctuations are required to drive the forward reaction i.e. crack propagation. In this situation growth rates will be high and governed by the ability of the stress field around the crack to adjust itself to translation of the crack tip. This rate of adjustment is controlled in turn by the inelastic properties of the solid.

Whilst there is no clear-cut transition between regime one and regime two, the following general comparisons are valid.

Regime I	*Regime II*
Low stresses	High stresses
Small flaws or none	Established cracks
Initiation phase	Propagation phase
Time/temp. dependence due to thermal activation of bond fracture	Time/temp. dependence due to rheological properties

Another feature that emerges here is the difference to be expected between brittle and extensible solids. For brittle materials, like glass, the transition from

Regime I to Regime II will be abrupt, since no rheological losses are available to provide time and temperature dependence of crack growth rate. For solids displaying plastic or visco-elastic energy losses the transition between thermally activated fracture and "stress-activated" fracture will be more gradual and perhaps not easily recognized.

A final consequence of this "reconciliation" between the kinetic theory and fracture mechanics concepts, is that the effective surface energy S should vary somewhat as Regime I gives way to Regime II. In the former, the stress field is required to provide energy (in the absence of losses) of the order of ΔG upwards, since the "balance" of the activation energy G^* is provided by thermal fluctuations. In Regime II virtually the whole of G^* must be provided by the stress field so that we have,

Regime I   Regime II
$\Delta G < S < G^*$ $S \sim G^*$

In practice it may not be possible to differentiate between these two situations since calculation shows that to obtain any appreciable rate of thermo-fluctuation breakage of co-valent bonds requires $S \sim 0.8\, G^*$ or greater even in Regime I [15].

1.4. Mathematical Formulation of Kinetic Theory

Once one accepts molecular fracture as an activated process, as outlined above, the theory of absolute reaction rates immediately leads to the following expression for the net rate v of the "forward" reaction A → B, i.e. net bond fracture.

$$v = \kappa \frac{kT}{h} \{[A] \exp(-G^*_{AB}/kT) - [B] \exp(-G^*_{BA}/kT)\} \tag{7}$$

where κ is a "transmission constant" k is Boltzmann's constant, h is Planck's constant, T is thermodynamic temperature and [A] and [B] are the concentrations of states A and B respectively.

When stress is applied to the specimen, the molecules themselves become stressed. Because the structure is non-uniform, the stress ψ on a given molecule will differ from the applied mean stress σ by a factor we may call s. Thus

$$\psi = s\sigma \tag{8}$$

$$0 \leq s \leq \psi_b/\sigma$$

where ψ_b is the stress under which the molecule will break without the aid of thermal energy. Some molecules will break without the aid of thermal energy. Some molecules will be overstressed ($s > 1$) and some will be understressed ($s < 1$).

It is clear that any tensile stress on the molecule will assist the process of bond fracture i.e. the activation energy barrier G^*_{AB} will be modified as previously indicated (Fig. 2). The rate constant for dissociation then becomes,

$$K_{AB} = \nu_o \exp\{-G^*_{AB} + f(\psi)\}/kT \qquad (9)$$

where $\nu_o = \kappa kT/h$ and f is a function. The precise form of f is not self evident, though a quadratic form would be expected for a fully extended molecule since the elastic energy stored is proportional to ψ^2. However, for molecular chains in the process of being straightened, a lower power would be expected since the effective modulus of such chains will increase as the chain approaches full extension i.e. will increase with stress. There is overiding experimental evidence from infra-red studies[16] that f is, in fact, linear with ψ, though the stress concentration factor s may itself vary with ψ.

We thus obtain

$$K_{AB} = \nu_o \exp\{-G^*_{AB} + s\beta\sigma\}/kT \qquad (10)$$

where β is a constant with the dimensions of volume (the "activation volume").

Although some theories (e.g. Knauss[17]) utilise the full expression of Eq. (7) to describe the forward reaction rate ν for chain fracture, it is usual to approximate this rate by

$$\nu = [A]K_{AB} \qquad (11)$$

The justification for this is that

(i) at levels of stress where significant stress-activated fracture occurs, the term in σ is sufficiently large to ensure that $K_{AB} \gg K_{BA}$;

(ii) due to retraction of the broken molecular ends, the effective concentration of states [B] available for recombination is reduced to zero; and

(iii) the primary radicals produced by fracture rapidly stabilise by combination with oxygen or by other decay processes, again reducing the population of reactive B states to a very low level.

Zhurkov and Tomashevsky[18] proposed a direct relationship between K_{AB} and the time to fracture of a loaded specimen. If we stipulate as a fracture criterion the requirement that a certain number N_c of molecular chains must fracture for the remaining intact chains to be unable to carry the load, the time to fracture, t_f, becomes,

$$t_f = \frac{N_c}{K_{AB}} = \frac{N}{\nu_o} \exp(G^*_{AB} - s\beta\sigma)/kT \qquad (12)$$

A plot of log t_f against applied stress σ should thus give a straight line of negative slope $s\beta/kT$, and this was confirmed experimentally for a range of polymers. The free energy of activation for bond rupture G^*_{AB} was also evaluated from these data and agreed closely with values obtained independently for the thermal rupture of chemical bonds[18].

In spite of this good agreement, Eq. (12) is somewhat naive in that it ignores the physical steps between bond fracture and macroscopic failure. These steps include the accumulation of fractured bonds to form microvoids or cracks, and the time dependent processes involved in crack propagation. The implied assumption in Eq. (12), therefore, is that the major part of the lifetime of a specimen under stress is controlled by the formation of a micro-cavity capable of propagating (rapidly) to give failure. As previously discussed, this is likely to hold good in Regime I, *i.e.* for relatively low stresses in elastic solids which do not contain pre-existing flaws.

The basic equations of the kinetic theory can be preserved in more sophisticated treatments of fracture which allow for some of the effects neglected by the simple theory.

One such theory[19] considers the statistical accumulation of molecular fracture events, in a plane perpendicular to the applied stress, for an assembly of chains under equal initial stress. Once a small number (say 3 or 4) adjacent chains have broken, the probability of further adjacent fractures increases rapidly and the process moves into a "thermal propagation" phase. Typical results as cited by Kausch and De Vries[20] are shown in Fig. 3.

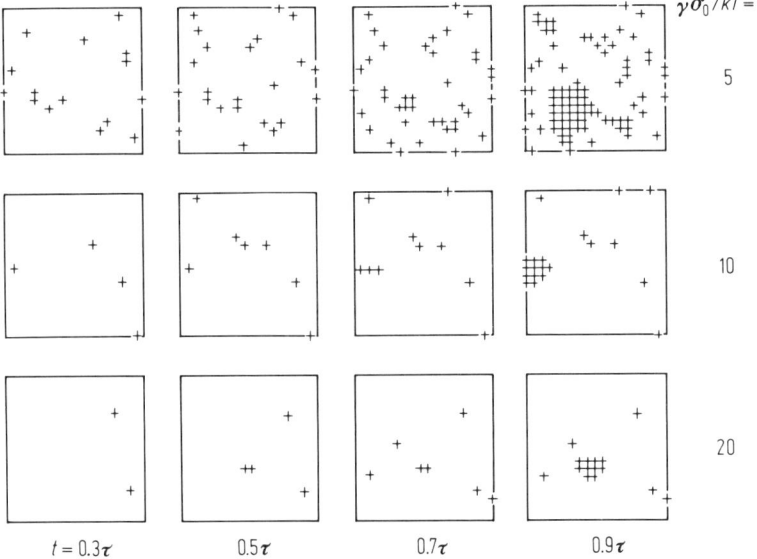

Fig. 3. The accumulation of statistical fracture events to form crack or void nuclei; γ and τ are constants of the system, σ_0 is the applied stress, t the time and T the temperature (after Ref.[20])

A second method of adapting Eq. (11) to a physically more realistic theory is due to Kausch and Becht[21] who considered the effect of chain length distribution on the accumulation of fractured chains. This theory is considered further in Section 3.7. and is able to explain both the time and strain dependence of molecular fracture events observed by ESR in stressed fibres. In particular it explains why, at constant specimen load, the rate of free radical production falls to zero after a certain time, in apparent contradiction to Eq. (1) which would imply a constant rate of molecular fracture under constant stress.

The kinetic theory can also be applied to crack propagation, using Eq. (10) to describe the fracture of molecules at the tip of the crack[17]. As already indicated, however, this approach has certain limitations in that it ignores the major role of inelastic processes. Referring back to Eq. (5), the kinetic theory can predict the variation of S with time and temperature in the transition zone between Regime I (thermally activated) and Regime II (mechanically dominated) fracture propagation. It does not offer any useful insight into the behaviour of the loss function Φ which largely controls time and temperature dependence of crack growth in Regime II.

2. Experimental Study of Molecular Fracture

Although the concepts of molecular fracture date back to the work of Taylor in 1947[22], it is only in recent years that techniques have been perfected for experimental study. Those techniques fall into three categories.

Firstly, there are indirect techniques in which the consequences of molecular fracture are assessed. They include a wide range of methods including the measurement of molecular weight changes and of solubility (gel formation). These have been extensively reviewed by Casale, Porter and Johnson[23]. Another such technique was employed by Becht and Fischer who observed the polymerisation of a monomer induced by molecular fracture in polymer chains[24].

Secondly there are direct techniques, notably electron spin resonance spectroscopy (ESR), in which the free radicals produced by the fracture of covalent bonds are directly observed, both in respect of their chemical nature and their number. Much of this review is concerned with the results of ESR studies and this technique is therefore treated at some length below. One little used technique for the direct assessment of free radicals produced by mechanical means is that of Pazonyi et al.[25] and Salloum and Eckert[26]. They chopped various polymers in an ethanolic solution of diphenyl picryl hydrazyl, a chemical indicator, and determined the free radical concentration in the cut surfaces by colorimetrie measurements of the colour change. This method is subject to some uncertainty on account of possible side reactions.

Finally, a number of associated techniques have been used to shed light on the phenomenon of molecular fracture. These techniques do not estimate the incidence of fracture events but rather concern phenomena or structures which bear strongly on the subject. Thus infra-red spectroscopy is employed to demonstrate the state of molecular stress prior to fracture, whilst low angle X-ray studies can be used both to assess the distribution of strain within a semi-crystalline polymer and to detect and measure the microvoids which are associated with molecular fracture in such materials.

In this Section we shall concern ourselves with the "direct" and "associated" techniques only.

2.1. Infra-red Spectroscopy

Two distinct types of study using IR have been carried out by Zhurkov and co-workers. The first technique[27] involves the determination of chemical end-groups in the polymer by measuring the intensities of their characteristic IR absorption peaks. After mechanical loading, the concentrations of various end-groups were found to increase significantly, suggesting that radicals produced by molecular fracture had reacted to form stable end groups. Some of the results of Zhurkov et al.[27] on polyethylene are given in the table below.

IR Peak frequency	Terminal group	Concentration x 10^{-18} cm^{-3}	
		Before loading	After loading
910 cm^{-1}	RHC=CH$_2$	18.8	25.6
1379	R–CH$_3$	40.0	60.0
1710	RCOOH	2.2	3.4
1735	RCOH	13.4	20.6
	TOTAL	74.4	109.6

The increase in end-group concentration, in this case some 35×10^{18} groups/cm^3, is assumed to equal twice the number of chain fracture events. It will be seen later that this number is about one to two orders of magnitude larger than the number of free radicals observed by ESR spectroscopy, though it has been argued[28] that ESR reveals only a fraction of the primary radicals actually formed. The IR estimate may therefore be a more accurate method of determining the absolute number of molecular fractures. Zhurkov et al. claim that the IR estimate agrees well with molecular weight changes measured on the same specimens after loading.

The second IR technique involves a quite different measurement, namely the shift in main chain vibrational frequencies induced by stress[16]. If ν_σ and ν_0 are the frequencies of the peak of a given IR absorption under stress and in the unstressed state respectively, Zhurkov et al. found that

$$\nu_\sigma = \nu_0 - \alpha \sigma \qquad (13)$$

where σ is the applied stress (see Fig. 4).

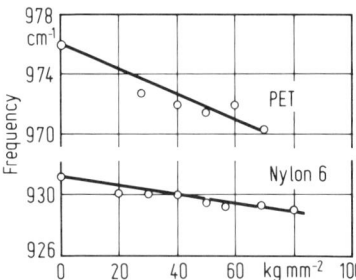

Fig. 4. Effect of stress on selected IR peak frequencies for two polymers (after Ref.[20])

This result supports the view that the stress modified activation energy barrier for bond fracture (see Section 1.) is linear in stress, since Bernstein[29] has proposed that the shift in dissociation energy D of a bond under stress is related to the corresponding shift in bond stretching frequency, thus,

$$c(D_o - D_\sigma) = \nu_o^2 - \nu_\sigma^2 \\
\simeq (\nu_o - \nu_\sigma) 2\nu_o \\
= 2\nu_o \alpha \sigma$$
(14)

Closer study of the absorption band shifted by stress showed a shape deformation which could be interpreted as a non-uniform stress distribution i.e. some bonds are more highly stressed than others and consequently suffer a higher-than-average frequency shift. Such bonds were designated "over-stressed" bonds and their distribution with respect to stress was deduced. An example is shown in Fig. 5 for

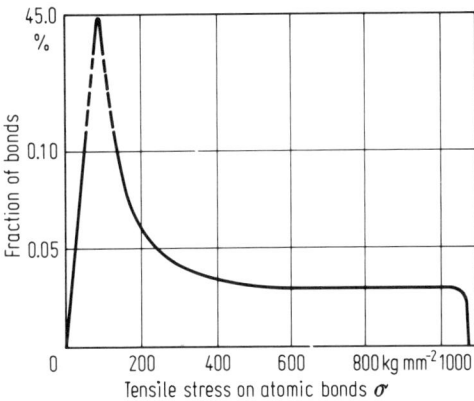

Fig. 5. Stress distribution over a bond population as determined by IR spectroscopy (after Ref. [16])

polypropylene under a load of 70 kg/mm². About 50% of the bonds are uniformly stressed, but some appear to carry stresses of up to 1000 kg/mm² (an s value of > 10). Similar results were obtained for PET and nylon 6. It was further shown, for PET oriented film, that the maximum stress on bonds is a function of temperature (Fig. 6) in harmony with the kinetic concept of bond rupture.

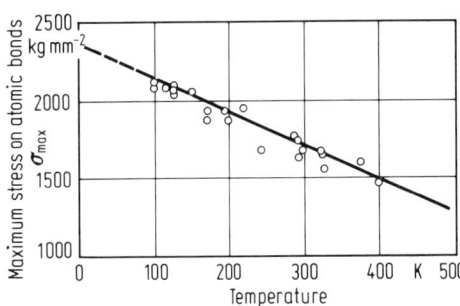

Fig. 6. Dependence of maximum bond stress in PET on temperature (after Ref. [16])

2.2. Low Angle X-ray Studies

Although low angle scattering is obtained from semi-crystalline polymers as a result of their crystalline-amorphous morphology, and can be used to measure interlamellar spacings, the scattered intensity from cavities or voids is very much stronger. It is therefore possible to use low-sensitivity data to follow the formation and growth of cavities under stress, using the Guinier formula[30]

$$I(\psi) \simeq ANV^2 \exp\{-(4\pi^2/3\lambda^2)L^2\psi^2\} \qquad (15)$$

where I is the diffracted intensity at angle ψ, L is the cavity dimension in the scattering direction, V is the volume of a scattering cavity, N is the volume concentration of cavities, λ is the wavelength and A a term related to cavity geometry and the electron density difference between cavity and solid. (For voids this electron density fluctuation is much greater than for crystalline-amorphous structures, hence the much greater intensities for voids.)

Using Cu Kα radiation (λ = 1.54 Å) and scattering angles from several minutes of arc to about 2 °C, cavity dimensions in the range 10 to 10^3 Å can be measured[14]. Figure 7 shows a typical set of scattering curves for nylon 6 (plotted as $\log I$ vs ψ^2) showing how the intensity increases with time under load. This is a result of the multiplication and growth of cavities with time.

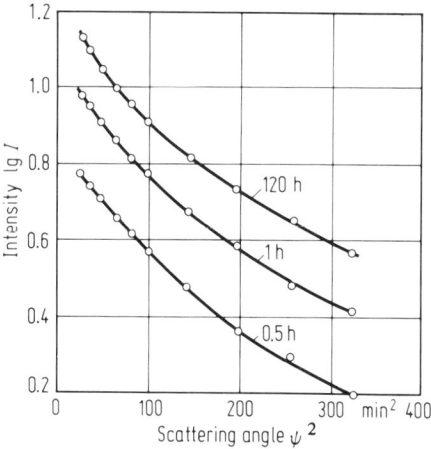

Fig. 7. LAXS intensity distributions for nylon 6 as a function of time (h) under load (after Ref. [14])

The curvature of the $\log I$ vs ψ^2 plots is a consequence of a size distribution, Eq. (15) being derived for an assembly of cavities of uniform size. If a suitable size distribution is chosen, the scattered intensity can be broken down into components which plot linearly and then the cavity size (obtained from the slope) and the number N of cavities (obtained from the intercept $\psi = 0$) can be deduced.

Using this analysis, Zhurkov, Kuksenko and Slutsker[14] followed the kinetics of cavity growth and correlated their results with ESR measurements of radical formation, as reported later in Section 3.

2.3. Electron Spin Resonance – Basic Concepts

Electron spin resonance (ESR) is one branch of spectroscopy, which studies the interaction between electronic magnetic moments and magnetic fields. The magnetic properties of the atom arise from the motion of the electron around the nucleus in orbitals and from the spin of the electron about its own axis. These motions give rise to orbital and spin magnetic moments respectively. A net magnetic moment exists in atoms or molecules containing unpaired electrons and only then does the ESR technique become possible. Application of an external magnetic field to a paramagnetic substance causes a splitting of the electronic energy levels, a phenomena known as the Zeeman effect.

It is not the purpose here to present a detailed account of ESR theory and technique. Many excellent texts cover the subject in considerable depth[31,32]. The interest here is the application of ESR to the study of molecular fracture in solid polymers. Generally the orbital and spin components of the magnetic moment are coupled, leading to a rather complex expression for the Zeeman energy levels. However, in solids, the orbital and spin motions become decoupled, and the electron then behaves similarly to a free electron. Consequently it is usual to consider only a paramagnetic species containing a single unpaired electron, for which the magnetic quantum number (M_j) takes the values $\pm 1/2$. In this case, the energy difference between the Zeeman energy levels takes the value

$$\Delta E = g\beta H_o \tag{16}$$

where g is the spectroscopic splitting factor, β the Bohr magneton and H_o the applied magnetic field. The population difference between the two Zeeman energy levels is given by

$$\Delta n_o = (1/2) N_o g\beta H_o / kT \tag{17}$$

where N_o is the number of unpaired electrons per unit mass. Hence ESR provides information on the number of radicals produced as a consequence of a particular process.

Electron spin resonance involving transitions between the Zeeman levels is achieved by applying electromagnetic radiation of frequency ω_o such that $\Delta E = h\omega_o$. The resonance condition is then satisfied

$$h\omega_o = g\beta H_o \tag{18}$$

where h is Planck's constant. The factor g takes a value close to 2 for the case of a free electron and h and β are true constants. In theory any combination of ω_o and H_o which satisfies the resonant condition is satisfactory. In practice ESR employs frequencies of 9–10 GHz with H_o in the region of 3200 gauss for $g = 2$. The oscillating magnetic field required to induce the transitions between the Zeeman energy levels is therefore in the microwave region. Since the radiation emitted during the transitions is polarized in a direction perpendicular to the main magnetic field (H_o), the oscillating magnetic field (H_1) is applied through a wave guide at right angles to the main field.

The main magnetic field is generated by large electromagnets. The dual requirements of strong main magnetic field, and the main field truly orthogonal to the oscillating field, severely restricts the volume of the zone for *in-situ* experiments and the sample size.

Energy is absorbed from the oscillating magnetic field (H_1) at resonance. Spectrometers available maintain a constant microwave frequency (ω_o) and vary the main magnetic field (H_o) to sweep through resonance. Such an absorption curve would appear as shown in Fig. 8(a). However, the net energy absorption arises from the difference in the upwards and downwards movements between the energy levels *i.e.* the energy absorption results from the population *difference* between the two energy levels which is small ($\simeq 0.07\%$). This problem is overcome by superimposing a small modulating oscillatory field (H_m) on the main field (H_o). The effect, as shown in Fig. 8(b), is that the crystal detector of the microwave bridge receives the first derivative of the absorption peak, rather than the direct absorption of energy. The a.c. output signal is readily amplified. The usual first derivative spectrum arising from the absorption curve shown in Fig. 8(a) is shown in Fig. 8(c) after phase sensitive detection. Most ESR spectra are presented as first derivative curves.

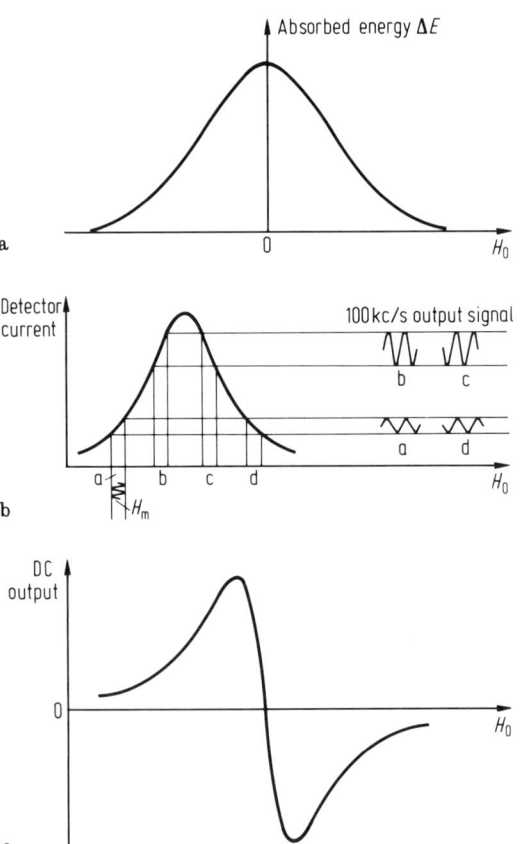

Fig. 8. a) Singlet absorption curve.
b) Superimposition of modulation on main field to produce phase sensitive a/c output corresponding to first derivative of absorption curve.
c) First derivative spectrum for singlet

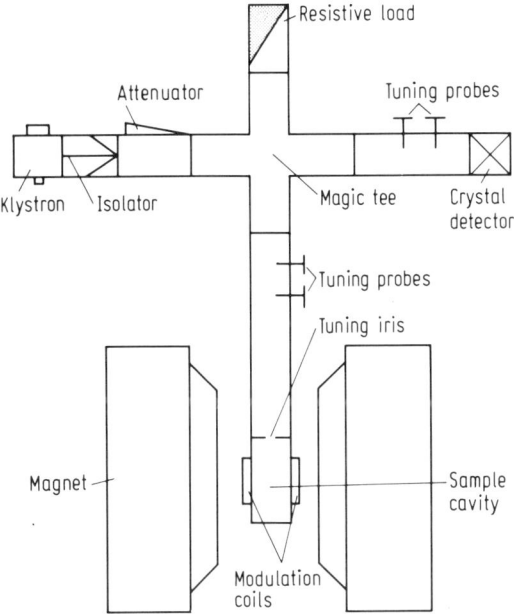

Fig. 9. Microwave components of one form of reflection-type ESR spectrometer

Figure 9 shows the schematic layout of a typical reflection type ESR spectrometer. Poole[32] provides an excellent commentary on many experimental aspects of ESR and ESR equipment. It is sufficient here to comment that the specimens are placed in the sample cavity, which forms part of the tuned microwave circuit. The sample cavity is of the order of 3 cm high, but the sample has to be confined to a narrow zone running through the centre of the cavity in which the magnetic field due to the microwave radiation is concentrated.

Reference to Fig. 8(a) shows that the absorption curve is of finite width over a span of H_0 and does not occur as a single line as the resonance condition would suggest [Eq. (18)]. The finite width of the absorption line is due to relaxation processes. Energy is absorbed in the transition upwards from the ground state ($M_s = -1/2$) to the upper state ($M_s = +1/2$). This energy must be dissipated to the surrounding material for the return to the ground state. Mechanisms by which this dissipation can occur are termed relaxation processes. Any mechanism which reduces the lifetime of the spin state causes a corresponding uncertainty in the energy content (Heisenberg uncertainty principle), which means that energy absorption from the oscillating field occurs over a range of frequencies (or a range of H_0). A careful study of line shape can provide valuable information on the relaxation processes operating. However such studies are not widely pursued with solid polymers.

Real systems exhibit line broadening for a number of different reasons, the most important of which are spin-spin interactions. Any particular spin experiences a magnetic field not only due to the applied main magnetic field, but also from magnetic dipoles of adjacent molecules. The resultant magnetic field induces resonance over a range of applied magnetic fields around H_0. Line broadening due to spin-spin interactions of adjacent molecules can be reduced by diluting the paramagnetic species, so that the species are separated. Such a remedy is not possible in solid polymers. Consequently line broadening frequently occurs, tending to obliterate the detail of the spectrum.

One very important interaction that occurs is that between spins within the same molecule, which results in hyperfine structure. In the case of polymers, it is the nuclear hyperfine structure which is important, which results from the interaction of the spin of the unpaired electron with nuclear spins from protons in the same molecule. Nuclear spin is quantized and for a nucleus of spin I, the electronic energy levels are further split into $2I + 1$ nuclear sub-levels. Figure 10(a) shows the energy levels resulting from a single electron spin ($S = 1/2$) and single nuclear spin ($I = 1/2$). The selection rule for transitions is $\Delta M_s = \pm 1$, $\Delta M_I = 0$, hence there are two values of main field (H_o) which satisfy the resonance condition and the selection rule. This gives rise to two absorption peaks and hence a two line spectrum, which is shown in Fig. 10(b). The maximum number of hyperfine lines is 2^n, where n is the number of interacting protons. If, however, all the protons are equal, the number of energy levels is reduced to $(n + 1)$, and the relative heights of the absorption peaks follow a binomial distribution.

An ESR spectrum is characterised by
a) the number of lines in the spectrum;
b) the position of the lines, usually in terms of the g values as computed from the resonance equation;

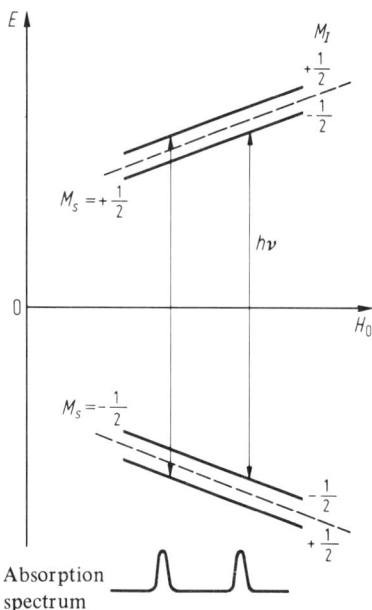

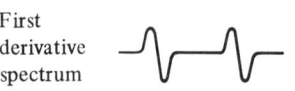

Fig. 10. Energy levels for a system with electron spin $s = 1/2$ and nuclear spin $I = 1/2$ as a function of magnetic field, together with absorption and first derivative spectra arising from the system

c) the separation of the lines (splitting) in terms of gauss;
d) the relative intensities of different absorption lines;
e) the line shape.

ESR spectroscopy was mainly developed to study the interaction energies of a parmagnetic atom in a constant magnetic field. It has therefore been mainly used to characterise rather than identify paramagnetic species. ESR has been used in polymer studies to identify the radical species generated by various processes, to determine the number of radicals produced and to follow the decay of radicals with time and environment.

2.4. Paramagnetic Centre Formation in Polymers

Since polymers are covalently bonded molecules formed from elements in the first two periods of the periodic table, all the electrons in the molecule are paired and there is no resultant magnetic moment. It is necessary to break covalent bonds to obtain unpaired electrons, before ESR spectroscopy can be applied. ESR thus provides information on the number and site of the broken bond.

Covalent bonds in polymers are broken during a number of processes, including thermal degradation, degradation by ultraviolet and γ-ray irradiation or by high energy particle bombardment. Much of the early work on the irradiation of common polymers has been reviewed by Campbell[33]. Irradiation is not highly selective in the choice of initial radical sites and free radicals may be formed either as a result of main chain cleavage, as in the case of poly(methyl methacrylate)[34–36] or by the removal of side groups, as in the case of polyethylene[37] and polypropylene[38].

More recently attention has been turned to the production of macroradicals during the mechanical destruction of polymers. Polymers have been ground, milled and crushed in a variety of ways, under inert atmospheres, vacuum or liquid nitrogen[39–41]. Free radicals are also produced in uniaxially stressed fibres, nylon fibres being commonly used for these studies[42,43]. A review and summary of the radicals identified from the mechanical degradation of polymers has been presented by Kausch[44]. Kausch tabulates the assignment of spectra of free radicals formed for a wide range of polymers for degration by milling and also those from stressed fibres and filaments, and Campbell reviewed the free radicals formed by irradition and Kausch those from mechanical degradation (see Section 4.).

2.5. Limitations on the Application of ESR to Polymer Studies

Since ESR can be used to monitor molecular rupture, it would seem to be a natural technique to use to monitor fracture at the molecular level in a wide range of mechanical testing events. However, it has been applied mainly to study molecular fracture during the rather extraordinary events of grinding or milling plastics and straining highly oriented fibres, often at temperatures around 77 K. De Vries, Roylance and Williams[45] sought to apply ESR to the impact fracture of PMMA, but were unable to obtain a spectrum, even though it was obvious that molecules were being ruptured in the process.

The three major limitations on the application of ESR to the general study of fracture in polymers concern,

a) the sensitivity of existing spectrometers,
b) the stability of the radicals formed,
c) the volume of material which can be placed in the cavity of the spectrometer, or the space available within the spectrometer for *in-situ* experiments.

The power absorbed at resonance, and hence the magnitude of the absorption peak is proportional to the population difference between the two Zeeman energy levels, which is typically only about 0.07% of the total population. Consequently a large total population is necessary to obtain a measurable signal. Present spectrometers have sensitivities which require minimum concentrations of $10^{11} - 10^{13}$ spins per gramme. While the size of the absorption peak can be increased by increasing the microwave power, this can lead to distortion of the spectrum. Consequently most studies on solid polymers have been carried out at low microwave power (typically $0.01 \rightarrow 0.1$ mW), which further reduces the sensitivity of the technique. To obtain manageable signals, it is therefore necessary to have a large spin concentration. This requirement is satisfied in irradiated samples, ground samples, (where the fracture surface area is large) and in tensile deformed, highly oriented fibres. Hence the extensive use of these techniques in the application of ESR to fracture studies. Backman and De Vries[46] ground and sliced samples of nylon 66, polyethylene and polypropylene at liquid nitrogen temperatures and found that the radical concentration increased proportionally to the fracture area formed. However, it was found that the number of broken bonds per unit area of surface formed was an order of magnitude less than the calculated number of molecules passing through the same area. An "easy path" fracture mechanism was suggested to explain this deficiency.

Free radicals formed by any technique are highly reactive and seek to react with surrounding matter to eliminate themselves. De Vries, Simonson and Williams[47] mention that the radical concentration is essentially independent of temperature up to approximately the glass transition temperature (T_g) of the material, but decreases rapidly above this temperature. Mead and Reed[15, 48] have measured the decay rate of radicals in tensile deformed oriented polybutadiene and found that the radicals were stable for several days at liquid nitrogen temperature, but decayed with increasing rapidity as the temperature was raised through the range 77 K to T_g. The decay rate curves are shown in Fig. 11. While the decay rate is very small at 77 K, at temperatures approaching T_g the radicals decay rapidly and decrease almost instantaneously around T_g. Since it requires considerable time to tune the microwave bridge once the specimen is placed in the cavity and further time to record the spectrum, it is essential to work with stable free radicals. Hence it is usual to work with polymers below their T_g and often at 77 K, using low temperature attachments on the ESR spectrometer. Since the population difference between the Zeeman energy levels is inversely proportional to temperature, it is beneficial to operate at low temperature from sensitivity considerations. However, relaxation processes may be ad-

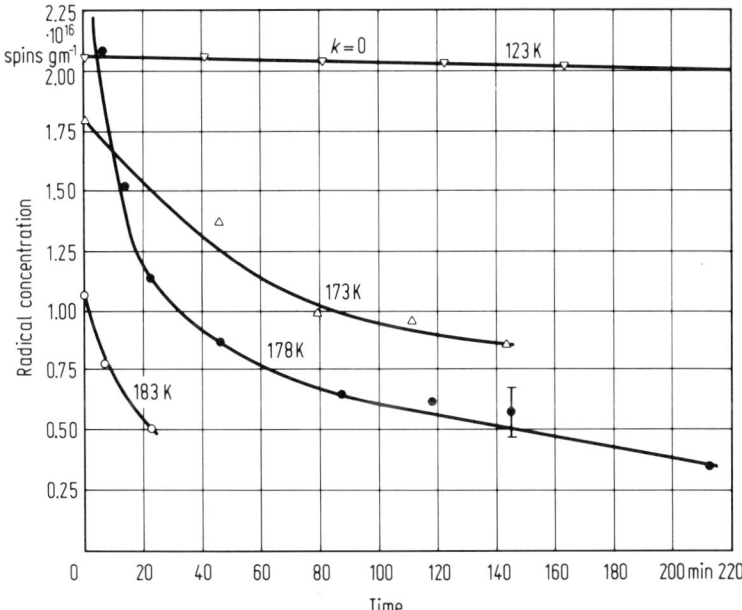

Fig. 11. Radical decay at four different temperatures in cis-1,4-polybutadiene after tensile testing at 83 K

versely affected, necessitating a reduction in microwave power to avoid line broadening and a consequential reduction in sensitivity.

The third limitation affecting the application of ESR to fracture studies is the size of the working zone within the spectrometer. Figure 12 shows a dimensioned sectional view of the standard cavity for a Varian E-9 spectrometer fitted with the low temperature attachment glassware. Low temperatures are obtained by passing a stream of nitrogen gas through a liquid nitrogen bath and over a thermostatically controlled heater before entering the sample zone. All glassware used in the cavity has to be spectroscopically pure. As can be seen from Fig. 12, the working zone is a mere 4 mm diameter by 20 mm. When the sample has to be placed in a tube, the diameter is reduced to 2 mm. Clearly the type of sample which can be accommodated in the standard cavity is limited. Liquids or powders containing a large radical concentration per unit volume are ideal, which again favours the grinding and milling techniques.

It is important to recognise that all free radicals within the cavity react with the magnetic field, and the spectra and radical concentrations determined are averages for the whole of the specimen within the cavity. ESR spectrometry cannot select areas for investigation as can, say, various types of microscopy. For example ESR has been used to monitor molecular fracture during crazing of polybutadiene at low temperatures[49]. While localised craze bands can be visually observed, the spin concentration can only be averaged over the whole length of the specimen. It is not therefore possible to determine the radical concentration locally around any particular craze band.

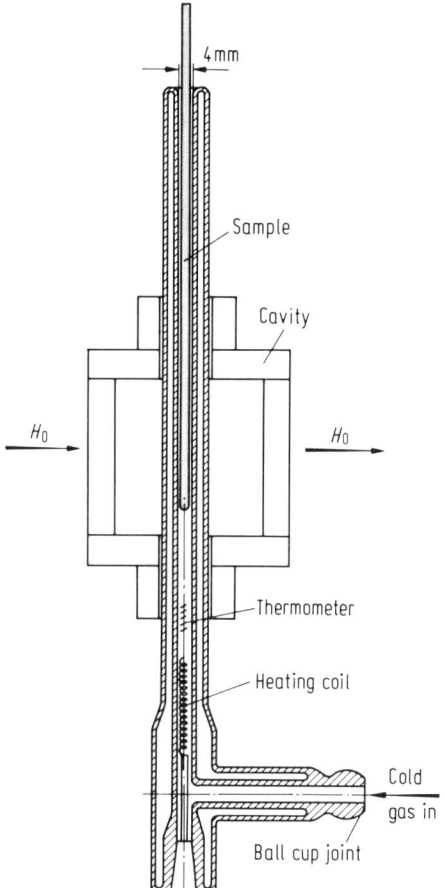

Fig. 12. Glass insert arrangement for low temperature operation in standard cavity of Varian E-9 spectrometer

2.6. Experimental Procedures

Irradiation studies are simply performed by introducing a suitable sample into an appropriately sized glass tube, before irradiating the sample at the desired temperature. The irradiated sample is then transferred rapidly to the spectrometer cavity, which may be pre-cooled to avoid decay of the free radicals. Some cavities are fitted with ports, so that the sample may be irradiated in-situ. The type of radiation required clearly limits the *in-situ* application. Since it is possible to introduce additional paramagnetic species by irradiation of the glass itself, precautions are usually adopted to avoid such artefacts. Similar empty tubes are irradiated for control purposes. Alternatively the tube can be sealed after introducing the sample, which is placed at one end for irradiation purposes. The tube is then inverted so that the irradiated sample falls to the "clean" end of the glass tube before performing the ESR examination. It is also common practice to evacuate the tubes prior to radical formation, since the radicals formed readily react with the environment, particularly oxygen, to form different radical species.

A range of *ad hoc* techniques are used for pulverizing the polymers in the glassy state. Specimens are normally prepared in specially manufactured glassware in which both the atmosphere and temperature may be controlled. The shavings or ground sample are then placed in a spectroscopically pure glass sample tube under controlled environment conditions and transferred to the ESR spectrometer. Details of one method are given by Backman and De Vries[46]. They developed a small electrically driven microtome to cut slivers from blocks of nylon, polyethylene and polypropylene. To control the environment, the whole device was placed in a double walled glass container; the space between the walls was evacuated. The atmosphere within the double walled container was nitrogen gas cooled to the required temperature. A jet of pure dry nitrogen was used to blow the slices cut into a collection tube, which was itself quenched in liquid nitrogen.

Backman and De Vries also filed nylon 66 submerged in liquid nitrogen using a small rotary file. Exact details of the method employed were not provided, as is often the case in papers on grinding and milling.

Small amounts of oxygen in the preparation atmosphere readily react with the radicals formed, even at liquid nitrogen temperature, and this necessitates its complete elimination at all stages. Backman and De Vries recognised this in their control of the preparation environment. A recent investigation on the effect of oxygen on the radical formation at low temperatures in polydiene rubbers has been carried out by Mead, Porter and Reed[50]. They have shown that the oxygen impurity in commercially available liquid nitrogen is sufficient to produce large quantities of peroxy radicals following molecular rupture at low temperatures. Hence oxygen-free nitrogen liquid and gas must be used for all stages of the specimen preparation. Samples have also been prepared by grinding in vacuum, but it is necessary to maintain the vacuum below 10^{-4} mm of mercury to exclude the oxygen[51].

Several devices have been developed for the *in-situ* tensile testing of fibres and filaments. To combat the poor sensitivity of the spectrometers, it is usual to strain several fibres simultaneously. Verma and Peterlin[42] looped nylon 6 fibres between two holders, so that the fibres passed through the cavity, with the holders outside the active zone. The holders formed sealed ends to "suprasil" quartz tube which could be evacuated to $<10^{-4}$ mm Hg. One holder was fixed, while the other holder was moved by a hand operated screw jack. No provision was made for simultaneously measuring load/deformation curves, which had to be obtained from conventional tensile tests on other fibres.

Reference to Fig. 12 soon indicates that there are considerable experimental problems in conducting *in-situ* tensile tests.

(i) The space available is limited,
(ii) It is necessary to have a straining frame which allows the specimen to pass through the cavity and be held at the top and bottom. The top-entry-only glassware shown in Fig. 12 is not suitable.
(iii) Any steel in the magnetic field surrounding the cavity distorts the lines of flux.
(iv) Provision must be made for *in-situ* testing either at controlled low temperatures in an inert gas stream or in high vacuum.

(v) Desirably it should be possible simultaneously to record load/extension date and ESR spectra, so that the two can be correlated. It is also desirable to be able to control the loading or extension input.

The straining frame used by Verma and Peterlin[42] satisfies some of these requirements, but control of the loading cycle was not possible because the straining frame was hand operated. A simple straining frame is described by Bates and Jackson[52]. This consists basically of a rigid G-shaped frame with a motor driven tensioning screw, operating through a gear box. Bates and Jackson apparently made no provision for environmental control or monitoring of the mechanical behaviour.

De Vries et al.[45, 53] give outline detail of a system they developed for a Varian E-3 spectrometer. A servo-hydraulic system was built which allowed loading of the samples in a wide variety of modes, and permitted tensile tests to be conducted at constant stress rate, at constant stress (creep), in cyclic fatigue, at constant strain rate, at constant strain or step strain. Provision was made for the simultaneous recording of stress-strain data and ESR spectra and a variable temperature control unit was employed.

A detailed account of a system developed for use with a Varian E-9 spectrometer is described by Mead and Reed[15, 54]. A wide access cylindrical cavity was employed, which permitted a working zone 25 mm in diameter and 50 mm long. Specially thin, vacuum insulated, double walled glassware was constructed for use with the wide access cavity, as shown in Fig. 13. The Varian low temperature control unit was used in a similar manner to that shown in Fig. 12 placing the heater and temperature sens-

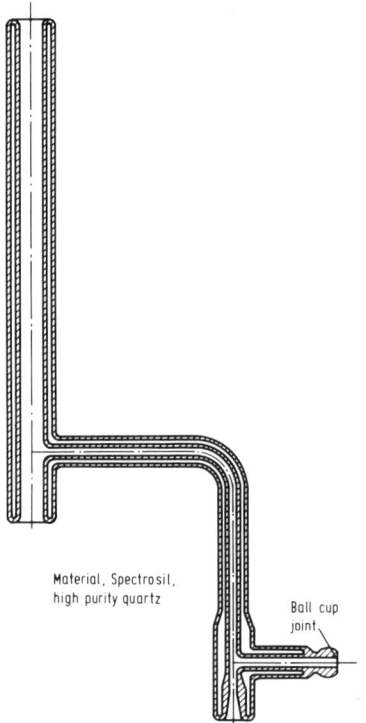

Fig. 13. Glassware used within ESR wide access cavity for insitu-tensile testing at low temperatures

ing element in the side limb. Cold nitrogen gas was caused to flow through the side limb to the wide bore test section within the cavity. The wide bore tube was open at both ends, so that the specimen could be stretched between clamps at either end. The lower clamp was attached to a load cell, the signal from which was fed through an amplifier to the Y-axis of an X–Y recorder. The upper clamp was driven at constant extension rate, through a system of gears by a servo-controlled motor. The servo-controlled system used ensured that a pre-determined extension rate was maintained, regardless of the load variations on the specimen. A displacement transducer attached to the upper clamp recorded the movement of the clamp, which was fed to the X-axis of the X–Y recorder. Extension rates in the range 0–50 cm/min. are possible with this apparatus. A straining frame was constructed around the magnet and cavity to guide the upper clamp and support the lower clamp and load cell. Special provision was made to raise and lower the entire straining frame, so that specimens could be mounted outside the glassware and cavity and then lowered into the test area.

A wide variety of techniques and apparatus have been developed for ESR studies of the mechanical degradation of polymers. These are mostly purpose built for a particular investigation.

3. Mechanical and Physical Processes

In this section we examine further some of the results obtained using the techniques reviewed in Section 2. Certain results and their implications have already been indicated for some of the experimental approaches discussed above, and here we shall concentrate on ESR data and the associated mechanical and physical phenomena.

The ultimate purpose of this discussion is to see how far macroscopic deformation and fracture phenomena can be interpreted in molecular terms. The techniques reviewed earlier, and especially that of ESR, provide direct evidence of molecular processes and we wish to see if this evidence is consistent with their dominant role implied earlier (in Section 1.) in determining macroscopic properties.

We shall see that the situation is far from simple and that the data on molecular processes cannot be used to make quantitative predictions about macroscopic deformation and failure properties. In particular it will become evident that the kinetic theory of fracture initiation is an oversimplification.

3.1. Factors Controlling Molecular Fracture

The approach taken in this section will be to consider in turn the empirical factors which govern the occurrence of molecular fracture and its consequences. The incidence of main-chain fracture in polymers under tensile stress is dependent on a wide range of variables, including polymer chemical structure, physical structure or morphology, additives, environment, time, temperature, orientation and physical state, as well as the more obvious variables of stress and strain.

It is helpful to think of main-chain fracture as a competitive response to deformation; if the deformation can be accommodated more readily by molecular flow or by crystallographic processes, then main-chain fracture will be less evident. This competitive situation underlies many of the phenomena reported in what follows.

3.2. The Importance of Molecular Anchorage

Polymers share with metals the general ability to undergo large deformations before fracturing. This is obvious in the case of elastomers but is equally important in tough engineering plastics which yield and flow under stress. These large deformations involve the uncoiling or straightening of the polymer molecules. In general, this is not a process which depends upon molecular fracture, though molecules may break as a consequence of such deformations. Exceptions to this statement can be found however, as we shall see.

On the other hand, molecular fracture becomes dominant if molecules are unable to flow because of structural constraints. These constraints may manifest themselves at the outset of a test or may emerge only after a significant deformation has already taken place, and after the molecules have been oriented along the line of action of the applied force.

Any structural feature which prevents mutual molecular flow can be regarded as a molecular anchorage. Experience shows that extensive molecular fracture (*i.e.* sufficient for detection by ESR spectrometry) can only be induced if such features are present. This does not mean that macroscopic fracture involves no molecular breakage in other cases. Main chain fracture limited to the tip zone of a single propagating crack, for example, would be undetectable by ESR. Without anchorages, however, it is impossible to induce *distributed* fracture of molecules thoughout a specimen under tensile stress which alone gives rise to detectable ESR signals.

Molecular anchorages are of two main types; crystalline regions and chemical cross-links (to which physical entanglements may also contribute in polymer networks). Not surprisingly, then, most ESR studies of polymers under tensile stress have been carried out using crystalline fibres (nylon, PET, PE, PP for example) or cross-linked polymers[20, 55, 48].

3.3. ESR Signals from Uncross-linked and Amorphous Polymers

Before considering data from crystalline and cross-linked polymers it should be emphasized that molecular fracture can be observed in such materials as glassy thermoplastics if these are ground or macerated[39–41]. By this means relatively large areas of fracture surface can be generated. Tensile tests on uncross-linked glasses, by contrast, generate only small areas of new surface and do not generally give detectable ESR signals.

Brittle fracture in glassy polymers occurs by the prior formation of crazes[56] which then fracture. In the craze, fibrils of highly oriented polymer are produced and it is the fracture of these fibrils that almost certainly produces the observed radicals. Even in this case, then, the signal derives ultimately from a fibre in which all potential for flow has been exhausted, leaving main-chain fracture as the only accessible deformation mechanism. It is possible that molecular fracture is also involved in the initial cavitation which gives rise to crazing, but this is much less certain for thermoplastic glasses.

Ironically, the results from grinding may have more direct relevance to macroscopic tensile fracture than the ESR data from tensile tests. This is because grinding does produce signals from fracture *surfaces* whereas tensile testing produces signals from distributable events which may or may not be associated with crack initiation and are unlikely to have *any* bearing on crack propagation. Unfortunately, grinding experiments are not susceptible of quantitative interpretation since neither the surface generated not the mechanical forces involved are accurately measured.

Indirect methods of assessing the incidence of molecular fracture in mechanochemical processes generally have been reviewed by Casale, Porter and Johnson[23]. One method, for example, is the determination of molecular weight changes in macerated materials. Direct measurements by ESR have been carried out by a number of investigators[46, 20]. Backman and De Vries, using several crystalline polymers cut up under liquid nitrogen, estimated that the average fracture surface involves breakage of between 5% and 10% of all molecules intersecting the surface. Salloum and Eckert[26], using drilled nylon, polyethylene and polypropylene, confirmed these results for nylon and polyethylene but obtained a much stronger ESR signal from polypropylene. But Pazonyi *et al.*, using an indicator to measure the radical concentration in plasticized PVC and PE cut under ethanol, obtained results which suggest breakage of between 34 and 120% of the chains intersecting a plane in the polymer[25]. Values in excess of 100% are not impossible since molecular fracture occurs not in a plane but in a layer of finite thickness (Backman and De Vries estimated a layer thickness as large as 6 mm within which 50% of the radicals were produced). Further work is clearly required to resolve this matter.

3.4. Crystalline Fibre Structures

The most extensive ESR studies of polymers under tensile load have undoubtedly been carried out on drawn crystalline fibres, and this work has been reviewed recently by Kausch and De Vries[20]. It is clear that the morphologies of oriented crystalline fibres, with extensive tie-chain populations and high degrees of molecular uncoiling, strongly favour the incidence of molecular fracture under tensile stress.

In spite of this, and somewhat surprisingly, some of the more obvious experiments have not been carried out. In particular there appear to be no data relating ESR intensities to the degree of crystallinity or other measurable variables by which morphology might be characterised.

Some data reflecting morphological effects has been obtained by annealing experiments on fibres of nylon 6 [57, 58].

Lloyd *et al.* found that the histogram of radical concentration versus strain was broadened considerably by annealing in a slack condition at 199 °C, and shifted towards higher strain by about 4% (Fig. 14). Interpretation is complicated by the fact that annealing produced a shrinkage of 20%, but the results do suggest that annealing reduces the homogeneity of the tie-chain length distribution.

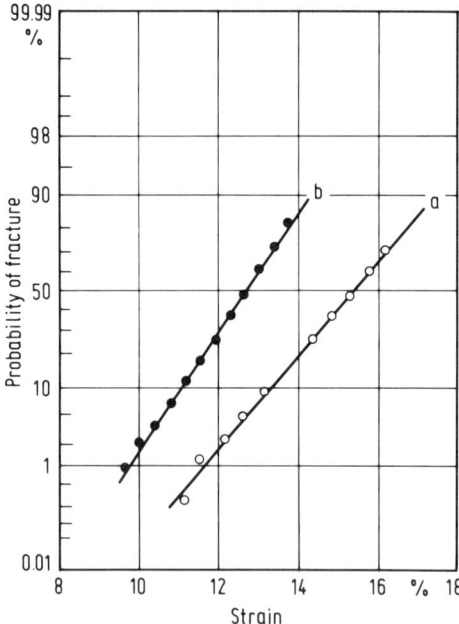

Fig. 14. The probability of fracture among a population of molecular chains as a function of tensile strain. Nylon 6. a) Annealed in a slack condition; b) unannealed reference specimen (after Ref. [57])

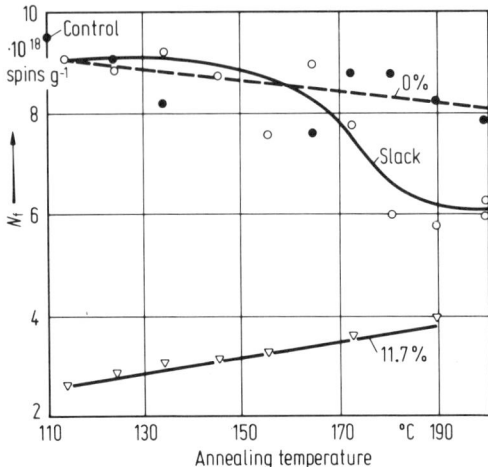

Fig. 15. Number of radicals at fracture in nylon 6 as a function of annealing temperature and annealing strain (after Ref. [58])

The maximum intensity of the ESR signal was also modified by annealing as shown in Fig. 15, the results being profoundly affected by the strain under which the specimens were held during annealing. If the maximum radical concentration is plotted against the observed fracture stress for the specimens, Fig. 16 results. The correlation between radical concentration and fracture stress for a given condition may indicate nothing more than that more molecules can be broken if the material

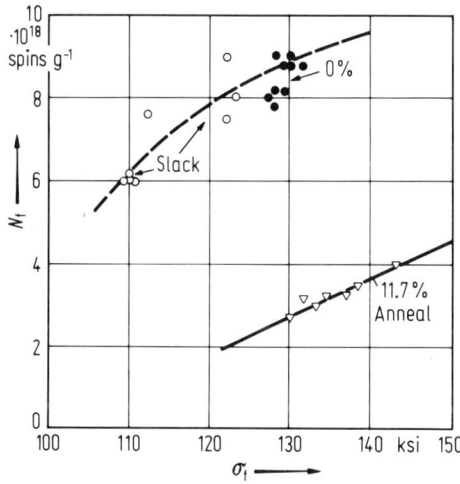

Fig. 16. Number of radicals at fracture as a function of tensile stress for different annealing conditions. Nylon 6. Open circles, slack; closed circles, 0% strain; crosses, 11.7% strain

sustains higher loads before fracture. The difference between specimens annealed at constant length (0% strain) or slack, and those annealed under 11.7% tensile strain is, however, more significant. It suggests that specimens annealed under positive strain suffer a loss of breakable tie molecules. This could be due to a number of causes, ranging from thermal fracture of up to three-quarters of the potentially breakable molecules to relaxations of such molecules to produce a 50% reduction in the *local* strain levels associated with a given test strain on the specimen. Whilst it is not possible to draw more precise conclusions, these data do indicate clearly that quite subtle morphological factors can greatly affect the concentration of radicals observed in tensile testing.

3.5. Glassy Polymers and the Role of Cross-links

Highly drawn crystalline fibres produce abundant radicals at strains exceeding about 8% and rising to the breaking strain of, say, 15 to 20%. Such fibres of course have already undergone draw ratios of 500% or more from the isotropic condition.

Polymeric glasses, in contrast, have to be deformed beyond the elastic region, *i.e.* beyond yield, before distributed molecular fracture occurs to a degree detectable by ESR. Cold drawing polycarbonate gave radical concentrations in the range 0.5 to 1.5×10^{16} spins/g, the concentration rising with strain rate in a way that suggests a direct dependence on the flow stress[59, 60].

No detectable signals can be obtained in the brittle fracture of isotropic polymers (even crystalline ones), but once plastic flow is induced (*e.g.* by small amounts of orientation above T_g before testing below T_g), radicals are readily obtained during deformation.

The radical concentration is strongly dependent on the presence of molecular anchorages, notably cross-links. Mead[15] studied cross-linked polybutadienes deformed plastically below T_g. Pre-strains of different magnitudes were applied above

T_g before testing, but comment here is limited to pre-strains below 250% where no strain induced crystallization occurs. Although weak signals ($\sim$ 0 to 8×10^{14} spins/g) were recorded in uncross linked material beyond yield, this radical concentration was two orders of magnitude smaller than that obtained on even lightly crosslinked specimens. Figure 17 shows the effect of cross-link density for specimens pre-extended by 200% on the radical concentration after a further 200% test strain and at various temperatures of test. The effects of temperature and strain will be discussed later.

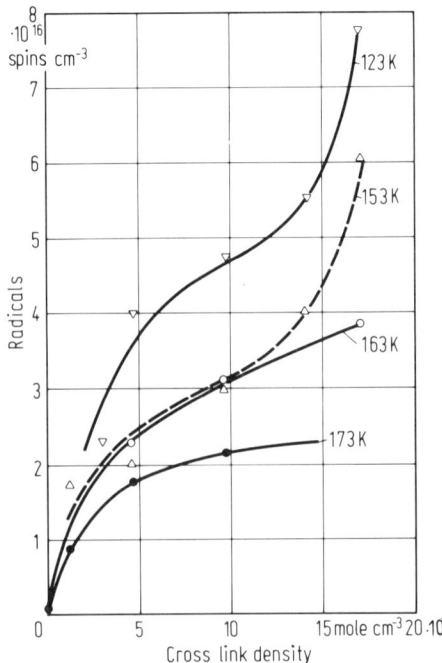

Fig. 17. Effect of temperature and cross-link density (expressed as mole/cm^3 of network chains) on radical production at 200% test strain after 200% pre-strain above T_g

The point to be noticed here is the major influence of the initial introduction of cross-links; raising the cross-linking from 0 to 10^{-4} moles/cm^3 of network chains produces a hundred fold increase in radical concentration whilst the further rise from 10^{-4} to 10^{-3} causes only a two to three fold enhancement of molecular fracture events. This suggests that the major role of cross-linking is to prevent total flow of a molecule relative to its neighbours. This requirement is achieved as soon as there are two cross-links to each molecule since this is sufficient to establish a continuous network. Another way to represent an incipient network, or gel point, would to be equate the length of a network chain to the average length of the original molecule and this would give a network chain concentration of roughly 2×10^{-5} moles/cm^3. This is just the range in which a rapid increase in radical formation is observed.

At lower temperatures (*i.e.* more than about 40 °C below T_g) a further upturn in radical formation is observed as the cross-link density rises above 10^{-3} moles/cm^3. This may reflect the higher stresses required to induce flow in the glassy material at these temperatures, suggesting that molecules are breaking, *before* they become fully straightened, as a consequence of inter-molecular friction.

A similar dependence of radical concentration on cross-link density was obtained for *cis*-polyisoprene tested below T_g by Natarajan and Reed[55].

3.6. The Role of Chemical Structure

The maximum radical concentration observed in cross-linked polybutadiene tested in tension below T_g is of the order of 4×10^{17} spins/g and for *cis*-polyisoprene, which may be more highly crystalline at large pre-extensions, of the order of 11×10^{17} spins/g. Chiang and Sibilia[61] have reported 1.5×10^{17} spins/g for PET and 0.76×10^{17} spins/g for nylon 6 at their ultimate elongation. Pazonyi[25] recorded a maximum of 8×10^{17} spins/cm³ for nylon 6 and Becht and Fischer obtained 10^{18} spins/cm³ for the same material at 17.4% strain. Other polymers commonly give significantly lower maximum radical concentrations, thus

PMMA, PS	$< 10^{14}$ spins/cm³	43, 18)
Polycarbonate	1.5×10^{16} spins/cm³	59, 60)
PE	5×10^{16} spins/cm³	18)
Nylon 66	10^{17} spins/cm³	62)

Natural silk[18] has a high value of 7×10^{17} spins/cm³. From these data it is clear that very similar amounts of molecular fracture can be obtained in polymers with widely different chemical structures, values approaching 10^{18} spins/g being recorded for *cis* PI, *cis* PBD, nylon 6 and natural silk under appropriate conditions. Secondly, the lower values obtained for other polymers seem to relate more to their morphology or state rather than their chemical structure (*e.g.* the very low figures for glassy polymers).

It appears, therefore, that chemical structure as such has little bearing on whether or not molecules can be induced to fracture under tensile stress. This is determined rather by the amount of strain that can be imposed before fracture, by temperature and by the physical and morphological characteristics of the material.

The chemical structure of the molecules does, of course, profoundly affect the *types* of radicals which are formed, and this subject is dealt with in the final section of this review.

3.7. The Roles of Stress and Strain

Our discussion of the kinetic theory of fracture in Section 1 has already indicated the manner in which applied stress can bring about a net accumulation of molecular breakages in a polymeric solid. Since the stress is continuous throughout a specimen loaded in tension, these breakages are distributed throughout the material and can be detected by ESR in terms of a volume concentration of free radicals.

Equation (11) represents the basic statement of the kinetic theory, namely

$$\nu = \nu_0 \, [A] \exp \left\{ -G^*_{AB} + s\beta\sigma \right\} / kT \tag{19}$$

where the applied stress σ accelerates the incidence of main chain fracture *via* the term $s\beta\sigma$, s being a stress concentration factor and β an activation volume. This equation describes the probability of fracture of a single bond. It can be applied to the rate of fracture of a population of bonds provided that the reverse reaction (*i.e.* re-combination) has a negligible probability. This will be true if free radicals are separated physically by molecular relaxation or are "neutralised" by reaction with oxygen or have a reverse activation energy G^*_{BA} (see Fig. 2) which is very large compared with the stress modified energy barrier $(G^*_{AB} - s\beta\sigma)$.

How far does Eq. (19) describe accurately the direct observations of ESR? The answer is not completely straightforward.

Equation (19) implies that, for small populations of broken molecules, the rate of radical formation, dN/dt, should be constant at constant stress. This was at first thought to be the case from the data obtained in constant loading-rate tests[18] but it was soon found[63] that stepwise loading tests told a different story. Figure 18 shows data of Becht and Fischer[24] for the stepwise loading of nylon 6. As previously observed by Roylance and De Vries, the rate of radical production under steady strain decreases to zero after a period of minutes (or even seconds), although stress relaxation is only of the order of 10%.

This dilemma is solved by considering in greater detail the actual morphology of the molecular network. Breakable molecular chains are considered to have a distribution of lengths so that the stress is not evenly distributed among load-bearing

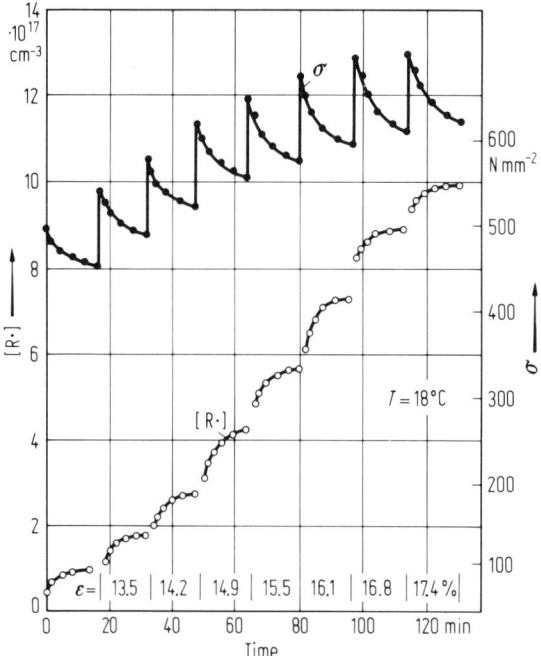

Fig. 18. Radical concentration and stress as functions of time and strain in step-strain tests on nylon 6 (after Ref. [24])

chains. We have already seen direct evidence of this from the infra-red studies of Zhurkov et al. At a given applied stress, the fraction of load-bearing chains under highest local stress do, indeed, fracture according to Eq. (19), but this results in a local re-distribution of stress which is more uniform than before. Although the average molecular stress increases as the shorter chains break, the *maximum* molecular stresses encountered are considerably reduced and the rate of fracture drops accordingly. Significant rates of new radical formation can only be restored by increasing the applied stress, thus re-populating the highest local stress states. The process is then repeated.

Kausch and Becht proposed that the local molecular stress could be written as[21],

$$\psi = E_R(\lambda_a L_o - L)/L \qquad (20)$$

where E_R is the elastic modulus of an extended chain, λ_a the extension ratio of an amorphous region, and L_o, L are respectively the end-to-end distance and the contour length of a stressed molecular chain. Using further basic equations relating the number of fractured bonds N to time and a rate constant K, thus

$$\left. \begin{array}{l} dN/dt = -NK \ (K \text{ allowed to vary with } t) \\[4pt] N(t) = N_o \exp\left(-\int_0^t K(\tau)d\tau\right) \\[4pt] Q(t_b) = 1 - N(t_b)/N \end{array} \right\} \qquad (21)$$

where Q is the cumulative distribution function of lifetimes amongst the chains, they obtained an expression for the radical concentration $[\dot{R}](t)$

$$[\dot{R}](t) = 2 \sum_i n_o(L_i) \{1 - \exp[-\int_0^t \bar{\omega}_b \exp(\beta \psi) d\tau]\} \qquad (22)$$

where i refers to the i-th fraction of chains (having length L_i) and $\bar{\omega}_b, \beta$ are kinetic parameters. They used a numerical iteration procedure to obtain a distribution of chain lengths L/L_o such that Eq. (22) fitted the data of Fig. 18. This distribution is shown in Fig. 19, where the fraction of breakable chains is plotted against $\log L/L_o$. The shortest chains represented by this distribution have $L/L_o = 1.06$ and the longest 1.18. The authors emphasize that the entire population of "breakable" chains represent only a small fraction of the total number of chains in the specimen, about 1.4% in fact. Furthermore the *load* carried by chains that break represents only about 20% of the breaking strength of the specimen.

The distribution of breakable chain lengths does not need to be evaluated using the rather complicated procedure of Eq. (22). A simple histogram of radical concentration as a function of specimen strain serves the same purpose as Fig. 19 also shows.

Histograms from stepwise straining tests, reflecting as they do the (apparent) length distribution of breakable molecules, are more informative than most other means of displaying ESR data. For example, Hassell[64] has obtained an interesting correlation between the width of the apparent chain length distribution and the

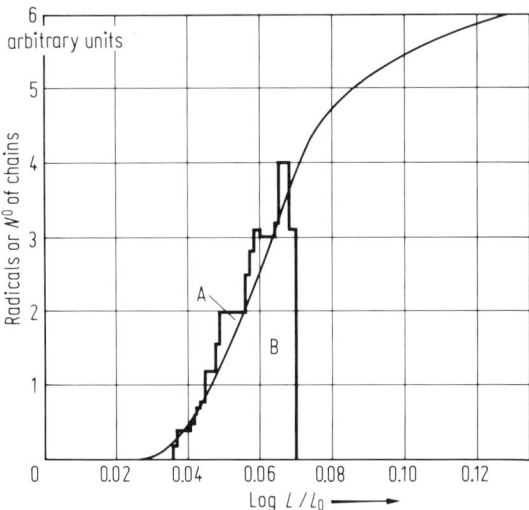

Fig. 19. (A) Distribution of tie chain lengths; (B) corresponding histogram for radical production (after Ref. [21])

ultimate tensile strength of the fibres concerned. This has not been interpreted[20] as a direct causal relationship because of the small concentration of molecular fractures relative to load bearing chains, but rather as reflecting differences in microstructure which itself controls macroscopic strength. A more direct interpretation should not, however, be ruled out.

3.8. The Effects of Pre-Strain and Plastic Strain

Highly oriented fibres produce ESR signals under tensile deformation in the ranges of 8 to 16% strain and 500 to 900 MN/m^2 stress. They do not undergo yield and plastic flow for the simple reason that the molecules are already highly aligned in manufacture and their capacity for plastic flow has already been exhausted. Such fibres therefore represent one extreme of the various conditions in which polymeric solids are obtainable.

Cross-linked elastomers (the other main class of polymers studied by ESR) can be pre-strained above T_g to any desired extent (up to fracture) and their orientation stabilised by cooling below T_g before testing. At high pre-strains, strain-induced crystallization may occur providing a morphology essentially similar to that of synthetic fibres[65]. Cross-linked polymers may therefore be used to explore in a systematic manner, the role of strain and orientation in molecular fracture.

The variables to be considered here are pre-orientation, test strain and temperature. The last named is also considered in a later section but cannot be omitted here because of its interaction with the strain variables. We must also consider the different categories of mechanical behaviour observed in these investigations.

Basically, four types of tensile stress-strain curve are found for cross-linked elastomers deformed below T_g. These are shown schematically in Fig. 20 where pre-

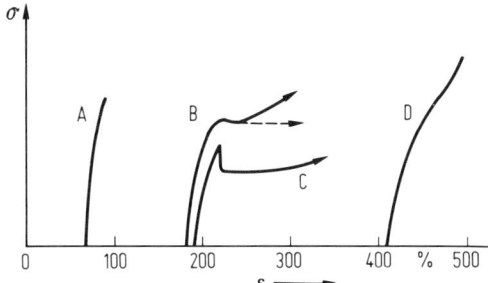

Fig. 20. Types of stress-strain behaviour observed in cross-linked elastomers below T_g. Origins of curves indicate amount of pre-extension (schematic)

orientation is indicated by the point on the abscissa at which the curve begins. For zero or low pre-orientation (curve A), brittle behaviour normally results and no detectable ESR signal is generated. At intermediate pre-orientations (say 100–300% according to cross-link density, temperature and strain-rate) two types of ductile behaviour can occur. Curve B shows yield and plastic flow accompanied by visible whitening of the specimen and the absorption of environmental gases. Here the primary mode of deformation is crazing which begins in distinct craze bands perpendicular to the strain axis but spreads progressively throughout the whole specimen. In contrast, "normal" ductile behaviour without crazing (curve C) is also found in this pre-orientation region. Figure 21 shows how the test temperature determines

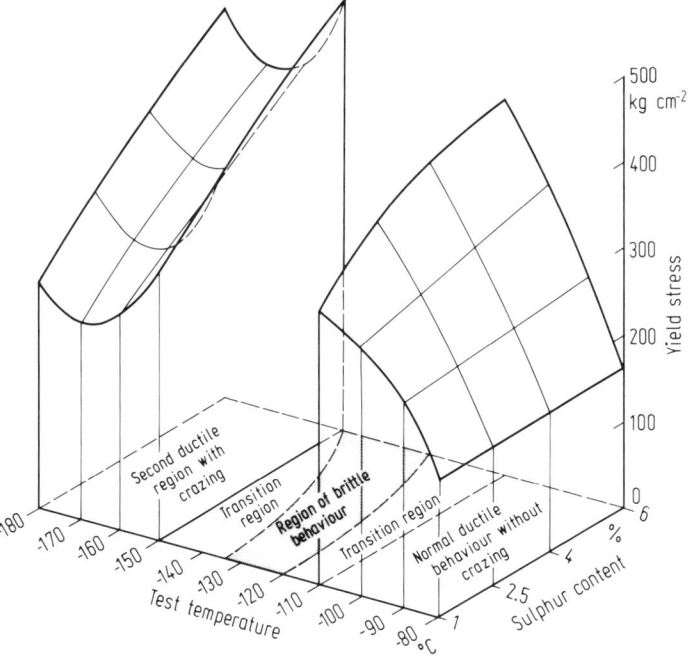

Fig. 21. Yield, crazing and fracture behaviour of *cis*-polyisoprene (cross-linked by sulphur) below T_g. Pre-orientation 100% (after Ref. [55])

whether crazing or non-crazing ductility occurs in 100% pre-oriented *cis*-polyisoprene. Normal behaviour without crazing, is observed at temperatures just below T_g and here the radical concentration is relatively low and short-lived[55]. (The decay of radicals at relatively high temperatures is discussed in the final section). A ductile-brittle transition occurs as temperature is further reduced but is followed by renewed ductility, now *with* crazing and strong, long-lived ESR signals.

The fourth type of stress-strain behaviour (curve D in Fig. 20) occurs with highly pre-oriented specimens. This behaviour approaches that of drawn fibres, though the crystallinity in polyisoprene and polybutadiene is considerably less than in, say, nylon 6. In this region strong ESR signals are generated by tensile testing below T_g but without visual evidence of crazing or voiding and without gas absorption from the environment. It is our view that the phenomena found with specimens of type D are qualitatively similar to those found in drawn crystalline fibres.

The effect of pre-orientation on the radical concentration at rupture in *cis* PB is illustrated by Fig. 22 [15]. It will be seen at once that the picture is complicated

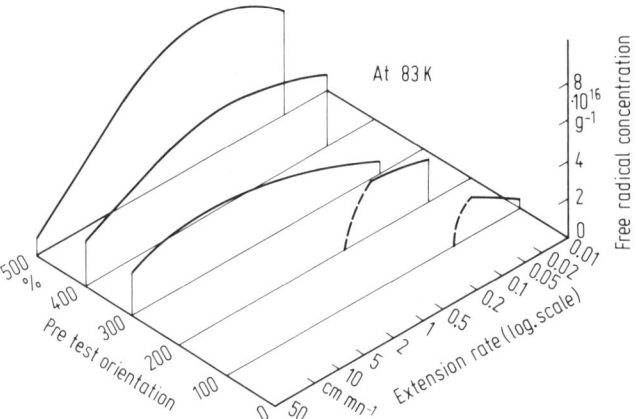

Fig. 22. Effects of pre-orientation and rate of testing on the maximum radical concentration observed in *cis*-polybutadiene tested at 83 *K*

by a strain-rate dependence. Thus whilst radical concentration generally increases with pre-orientation as expected, at the highest strain rates it actually *decreases* with pre-orientation and this occurs even though the fracture strain is still rising with increasing pre-orientation. These effects can be explained qualitatively by supposing that higher pre-orientations introduce more molecular chains extended to near their breaking point and so, generally, causes an increase in molecular fracture events. If, however, such fracture takes place, not spontaneously, but by a mechanism of micro-void growth (see later) the time for such growth may not be sufficient at high strain rates to effect the rupture of all the potentially breakable chains. This accounts for a decrease in radical formation with increasing strain rate. The decrease with increasing pre-orientation at high rates requires a further supposition that micro-void formation is delayed in highly pre-oriented materials. This may be because the molecular stress distribution is more uniform as a result of shorter chains having already fractured during pre-orientation above T_g (evidence of network breakdown in highly strained cross-linked systems is considerable[66]).

The effect of increasing test strain on radical concentration is evident. Figure 18 for crystalline fibres has already indicated an almost linear dependence of radical concentration on test strain between 10 and 20%. This is plotted in Fig. 23, where the slope $dN/d\epsilon$ is 24.10^{18} spins/cm^3. In the plastic deformation of cross-linked glasses, where the test strains are of course very much larger, roughly linear plots are again observed though there is a consistent increase of $dN/d\epsilon$ with increasing strain.

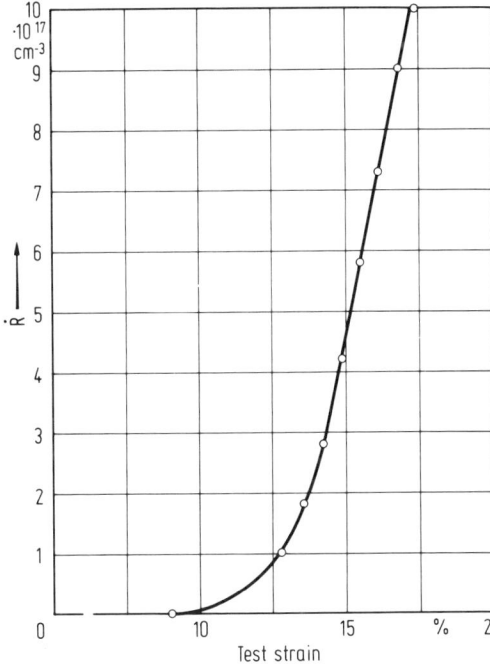

Fig. 23. Radical concentration as a function of test strain in Nylon 6 at room temperature (data replotted from Ref.[24])

The measured values for $dN/d\epsilon$ for plastic deformation of cis-polybutadiene[15] range from about 8.10^{17} for a pre-extension of 400%, a low temperature and a high cross-link density (1.7×10^{-3} moles cm^{-3}) to zero for low pre-orientations and higher temperatures. The effect of temperature is illustrated in Fig. 17.

For plastic flow without significant strain hardening the value of $dN/d\epsilon$ is two to three orders of magnitude smaller than that for drawn fibres and only begins to approach the latter value at the highest pre-orientations. It is clear, then, that the efficiency of strain in causing molecular fracture is highly dependent on the degree of initial molecular orientation and this can be interpreted in terms of the population of nearly-extended chains at the outset of the test. The effect of cross-link density and the increase in $dN/d\epsilon$ as plastic flow proceeds to high levels of strain, are consistent with this view. It should also be noted that breaking stresses in the crys-

3.9. The Effects of Strain Rate

The effect of straining rate on maximum radical production in drawn fibres is not great, as Fig. 24 shows[67], although we have already noted that under steady load the number of radicals increases with time for periods of up to twenty minutes. In the plastic deformation of polycarbonate also, radical concentration increased only two-fold over two decades of strain rate[59, 60]. Much larger effects have been observed in plastically deformed, cross-linked glasses, as Fig. 22 shows. Here, for *cis*-polybutadiene[15] at 500% pre-strain, the radical concentration at fracture has a peak value of 7×10^{16} spins/g at intermediate strain rates, and decreases to 5×10^{16} at

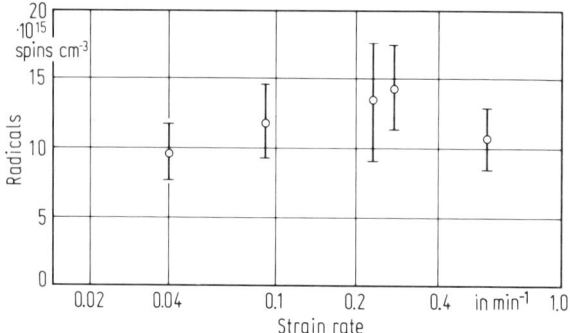

Fig. 24. Effect of strain rate on radical concentration in nylon 6 fibres (after Ref. [67])

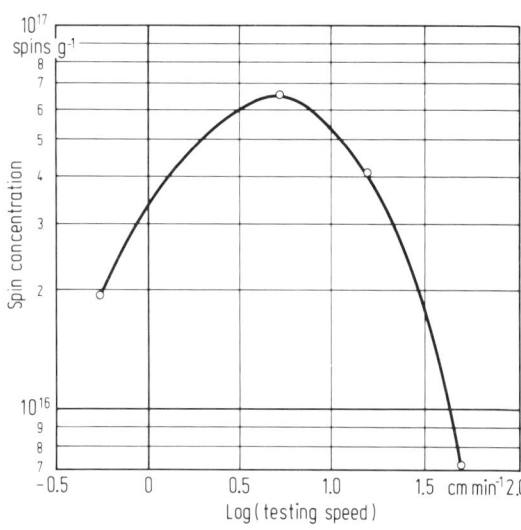

Fig. 25. Effect of testing speed on radical concentration in polychloroprene at 123 K. Pre-orientation 300% (after Ref. [68])

the lowest and 1×10^{16} at the highest strain rates employed (a range of nearly four decades). Even more dramatic effects were obtained[68] for radical concentration in cross-linked polychloroprene (below T_g) at a fixed test strain (Fig. 25). In this case the number of radicals formed seemed to correlate with the amount of visual crazing. It is likely that the relatively strong influence of strain rate in plastically deforming specimens is closely related to the growth of crazes or sub-microscopic voids. At very low strain rates molecular fracture may be reduced by molecular flow. As the rate increases, so also does the flow or yield stress giving rise to a higher incidence of fracture. As strain rate further increases, however, crazing is suppressed (see Fig. 22) because the necessary time for extensive void growth is not available and molecular fracture associated with such growth is progressively reduced.

3.10. The Effects of Temperature

The kinetic theory equation for bond breakage probability v, namely

$$v = v_o \exp \{-G^* + \beta \psi\}/kT \tag{23}$$

obviously suggests that radical formation should depend strongly on the temperature of test. Several workers have demonstrated such effects in the tensile loading of drawn fibres[63, 69]. Roylance et al. showed that the steady state radical concentration N_{ss} should be related to the applied stress σ by,

$$N_{ss} = \delta \exp \xi \sigma \tag{24}$$

where $\delta = \alpha \exp(-\gamma/kT)$ and $\xi = \beta'/kT$, α, β' and γ being constants for the material. Plots of ξ vs T^{-1} and of $\log \delta$ vs T^{-1} for drawn nylon 6 fibres (Fig. 26) confirmed the predicted temperature dependencies.

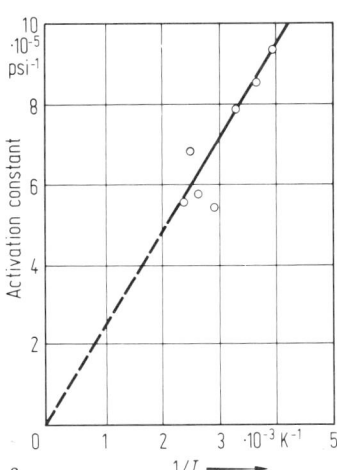
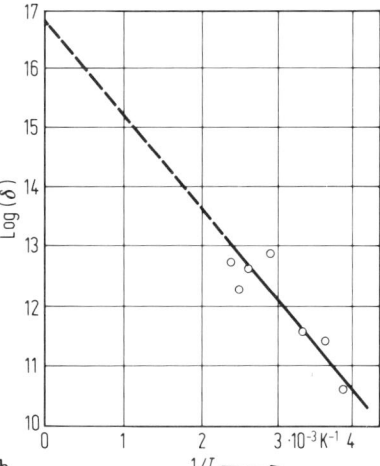

Fig. 26. Effect of temperature on a) the activation constant ξ and b) the pre-exponential constant δ, of Eq. (24) (after Ref.[63])

Johnson and Klinkenberg carried out step-temperature changes on strained polyamide fibres and found the expected increase in radical concentration with increasing temperature (Fig. 27). The temperature affects not only the total radical production but also the apparent distribution of breakable chain lengths, as the histograms of Fig. 28 show. It appears that testing at higher temperature permits higher breaking strains and thus a broader distribution of chain lengths available for rupture. At the same time the breakable chain population at a given length is greatly diminished. This may be due partly to accelerated radical decay at higher temperatures (see later) which may also offset the expected rise in radical concentration with increasing temperature.

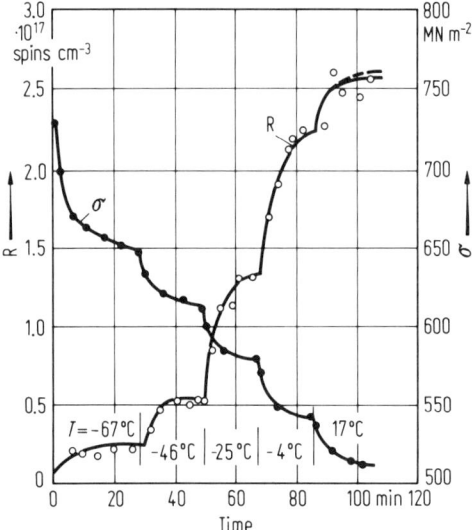

Fig. 27. Spin concentration and stress as functions of time and temperature during step-temperature change tests on nylon 6 fibres at constant strain of 14.2% (after Ref. [69])

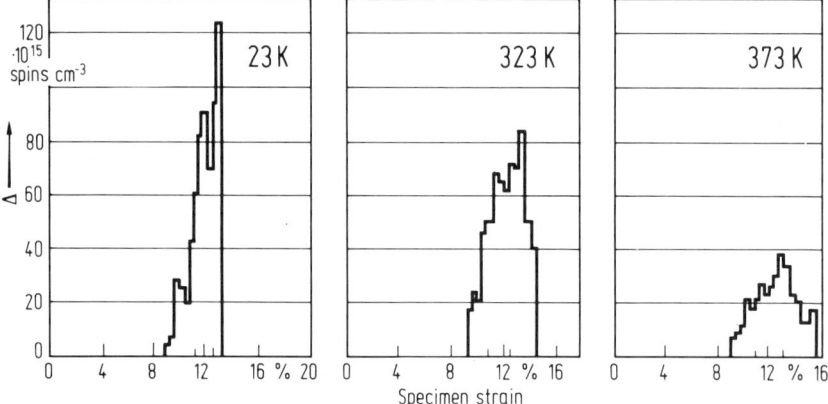

Fig. 28. Histograms of spin increments against strain for nylon 6 fibres, showing effects of temperature (after Ref. [57])

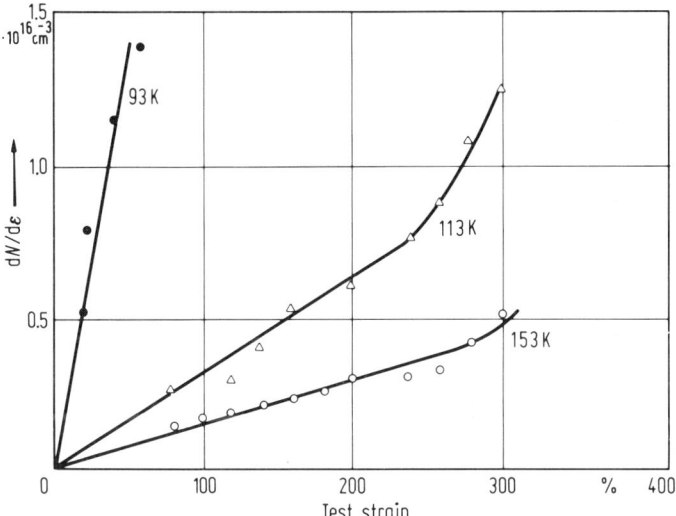

Fig. 29. Effect of temperature of test on the rate of radical formation, $dN/d\epsilon$, versus strain curve for cross-linked *cis*-polybutadiene. Pre-extension 200% (after Ref. [15])

The results on plastic deformation of cross-linked glasses appear at first to contradict those discussed above for crystalline fibres. Figure 29 shows the effect of temperature on the plot of radical concentration versus strain for *cis*-polybutadiene[15]. The number of radicals increases with decreasing temperature, instead of decreasing as might be expected. This effect may be attributable to radical decay during test, but other evidence indicates that secondary radicals at least are long lived below 123 K. The more likely explanation is that the flow or yield stress increases rapidly as the temperature falls, so that at 93 K the molecules are subject to much higher shear stresses than at 123 K. The ratio of radical concentration to yield stress appears to decrease with decreasing temperature in harmony with the expectations of the kinetic theory.

3.11. Crazes and Microvoids

Our final topic in this Section concerns the physical process of void growth which is so often associated with the observation of ESR signals in the tensile testing of polymers. We shall also refer at this point to the important effects of the environment on these processes. This topic is particularly important since it is only as we consider the growth of cavities that our discussion of *distributed* molecular fracture converges with the subject of macroscopic fracture with which we began this review.

Zhurkov et al.[14] used light scattering and low angle X-ray scattering (LAXS) to study the time-dependent formation of microcracks or cavities in stressed polymers. The concentrations of microcavities were found to be as high as $10^{15} - 10^{16}$ cm^{-3}, uniformly distributed throughout the specimen, though more numerous close to the specimen surface.

Figure 30 shows how the number of microvoids increased with time under load in a PVC film, though a similar behaviour was found in oriented polycaproamide (nylon 6). The initial rate of formation of voids, dN/dt was logarithmic with tensile stress. A further important observation was that the critical concentration of cavities just before macroscopic failure appeared to be constant for specimens whose lifetimes under load varied over four decades of time (according to the applied load). This is shown in Fig. 31. The same critical concentration of microvoids was found close to the tip of a propagating crack.

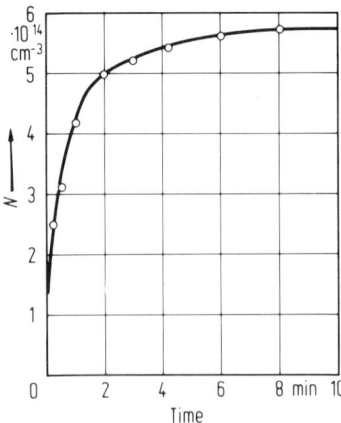

Fig. 30. Growth in number of microvoids with time in PVC under a tensile stress of 6 kg/mm^2 (after Ref.[14])

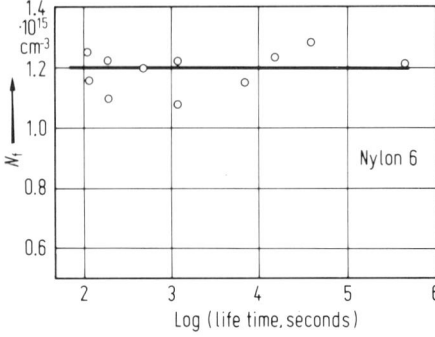

Fig. 31. Concentration of microvoids at failure as a function of time to failure for nylon 6 fibres (after Ref.[14])

The microvoid dimensions were such that the ratio of size (L) to their mean separation (R) was between 0.25 and 0.5. The cavities appeared to be disc-shaped with their planes perpendicular to the stress axis and had an average size around 300 Å. "Small" and "large" microvoids could be recognised in the scattering patterns and whilst most were small, the number of "large" voids increased with time.

In further studies Zhurkov et al.[70] measured microvoid formation and bond rupture on the same specimens. Bond fracture was assessed using the infra-red technique described in Section 2. They obtained identical behaviour for the kinetics of bond rupture and cavity accumulation, both in respect of time and stress dependence. Their results are summarized in the following table:

Polymer	F, cm^{-3}	N, cm^{-3}	D, Å	R_1	R_2
Nylon 6	27×10^{18}	900×10^{14}	90	300	330
p-Propylene	1.4×10^{18}	5×10^{14}	320	2800	2600
p-Ethylene	6×10^{18}	6×10^{14}	400	10000	11000

Here F is the bond rupture concentration, N the microvoid concentration and D the diameter of the disc-shaped microvoid. The two ratios R_1 and R_2 are respectively the number of molecular fracture events per cavity (F/N) and the number of aligned molecules calculated to intersect a plane of the same area as a microvoid. The close agreement was taken to signify that all molecules passing through a microvoid area had been fractured by the growth of the void.

This is a reasonable conclusion for highly oriented crystalline polymers where the amorphous regions are probably composed largely of tie molecules. It is unlikely to be true of microvoids and craze cavities formed in amorphous glasses or poorly oriented crystalline polymers.

The occurrence of crazing in the plastic deformation of cross-linked glasses has already been referred to. Visible crazing in polybutadiene and polyisoprene was obtained under some experimental conditions but not all. For crazing to occur required, in general, low strain rates, low temperatures and low pre-orientations (though not so low that brittle behaviour ensued). A feature of visible crazing is the absorption of relatively large quantities of the environmental gases. The quantities are too great to allow their presence in vapour form within the voids[48] and it follows that they must be present in a condensed form, possibly as a monolayer within the craze.

This observation, along with the work of Brown[71] leads to the conclusion that the crazing observed is environmentally induced, and occurs only at temperatures just below or just above the boiling point of the environmental gases. This hypothesis was confirmed by Mead[15] who deformed polychloroprene and polybutadiene at various temperatures in atmospheres of nitrogen, helium, carbon dioxide and sulphur dioxide. Mead also found corresponding changes in the load-extension curves of his materials for different environments. When conditions were not suitable for environmental crazing, mechanical behaviour reverted to either brittle fracture or shear yielding.

Brown et al.[71] have shown that, in PMMA, both environmental and non-environmental ("intrinsic") crazing can occur, and have shown how the incidence of the former depends on the partial pressure of the crazing agent. It is not certain whether intrinsic crazing occurs in the low temperature deformation of cross-linked glasses or whether craze formation is wholly environmental in these cases.

It is not at all certain that the void formation and crazing discussed in the previous paragraph is directly associated with molecular fracture in the manner proposed by Zhurkov for crystalline fibres and films. ESR signals are certainly produced simultaneously with visible crazing in cross-linked glasses, and the chemical nature of the radicals observed is affected by the penetration of environmental gasses. Furthermore, radicals are not always detected in the shear yielding of these materials. On the other hand, even stronger ESR signals are obtained in those deformation regimes (*e.g.* for polybutadiene) where no visible crazing is observed. It should be remembered, however, that Zhurkov's microvoids do not appear as visible crazes.

In spite of the caution necessary in interpreting the data, it is not unreasonable to suppose that crazing, like microvoid formation, does involve molecular fracture at least in cross-linked glasses. This could occur both in the initial cavitation event and in the subsequent growth of craze cavities, the latter involving plastic flow in the "walls" of the cavity. It is unlikely that a one-to-one relationship exists between molecular intersections and void area in crazing, though Zhurkov may be correct in proposing such a relationship for oriented crystalline polymers. Finally, microvoiding may well be taking place in the more highly oriented cross-linked glasses where visible crazing is no longer evident.

The enhancement, or even inducement, of crazing by environmental gases is almost certainly a physical effect rather than a chemical one[72]. That is, it is most unlikely that this process affects the incidence of molecular fracture. What the adsorption of environmental fluid may well do, however, is to make possible the growth of microvoids into visible crazes. The reason for this is that the energetically unfavourable step in the cavitation of a continuous medium is the expansion of the cavity at very small radius against the surface tension of the medium. (The stress required goes as $2\gamma/r$ where γ is surface tension and r is radius). Adsorption, by reducing γ, favours cavity growth at this critical stage and thus greatly improves the chances of large cavities developing.

3.12. Synopsis of Mechanical and Physical Processes

Finally in this section we attempt to draw together the evidence provided by the various techniques employed to study molecular fracture. For clarity, this synopsis is presented as a brief catalogue of sequential events linking the application of stress to macroscopic deformation and fracture *via* the molecular processes that may occur.

a) The application of tensile load to a solid polymer elicits various responses at the molecular level depending on the morphology and state of the solid. Molecules which are randomly coiled tend to uncoil and straighten in the direction of stress. If the material is glassy, this can only occur at high stresses (*i.e.* above the yield stress) because of the steric hindrance of neighbouring molecules. Molecular chains which are already fully extended (by pre-orientation or fibre formation) will deform elastically to accommodate the load. Some which are close to a full extended configuration will be brought to full extension as a result of loading.

b) Molecular chains which are fully extended, being less compliant than their coiled neighbours, will carry a disproportionate fraction of the total load, as in any "composite" solid. The high stresses on the covalent bonds of such a molecule may by detectable by shifts in the associated vibrational (infra-red) frequencies. Molecular regions in crystals will share load uniformly and are unlikely to become "overstressed".

c) Because of the high dissociation energies of co-valent bonds, intermolecular slippage or flow, involving the rupture of secondary bonds, will normally occur in preference to molecular fracture (*i.e.* main-chain breakage). Under such circumstances, plastic deformation by flow will be observed without molecular rupture. This flow may be uniform as in shear yielding or localised as in void growth and crazing.

d) If the molecules are "anchored" in the structure either by crystalline regions or chemical cross-links (or even by persistent physical entanglement with *e.g.* long relaxation times) the mutual flow of such molecules will be prevented. They will thus uncoil as the strain increases and eventually become fully extended and elastically loaded. This may occur after small strains (as in oriented fibres) or after large test strains (as in plastically deforming glasses). As already noted, non-uniformities of structure will ensure that different molecular chains achieve full orientation before others, so that the population of fully-extended chains will be strain, and pre-strain, dependent. The degree of structural non-uniformity will depend upon morphology and thus upon thermal treatments and other factors.

e) In the absence of permanent molecular anchorages it is still possible for individual molecules to become highly stressed. This may occur, *e.g.* under the high stress necessary to produce flow at temperatures far below the glass transition. A long molecule may have physical entanglements with its neighbours which relax on a time scale long compared with the deformation rate and thus act as "permanent" anchorages on this time scale.

f) Fully extended molecules fracture under the combined influence of mechanical stress ("mechanical excitation") and thermal fluctuations, according to Eq. (19) and the kinetic theory of fracture. The rate of this process is governed by the molecular stress and the temperature.

g) As a result of the fracture of fully extended chains, stress is redistributed more evenly and the rate of chain fracture falls at constant overall stress. The process of mechanical degradation can only progress, therefore, if the applied stress or strain is increased to re-populate the fully-extended state.

h) Because stress is redistributed by molecular fracture *locally*, such fracture events tend to accumulate locally, producing a micro-void.

i) Micro-voids, which are distributed uniformly throughout a specimen, can grow by various methods:
(i) by propagation as a result of further molecular fracture
(ii) by growth as a result of local molecular flow without fracture, possibly aided by environmental adsorption reducing interfacial energies, and
(iii) by coalescence.

j) The growth of microvoids can result in either a single large local flaw or crack which propagates to destroy the specimen, or else in a stable array of voids as in the wholesale crazing sometimes observed. The conversion of a narrow craze, by craze breakdown, into a propagating crack is an intermediate possibility.

k) The establishment, or nucleation, of a single crack is only the first step towards fracture. For macroscopic failure to be observed, the crack must propagate through the specimen. Propagation involves both the creation of new surfaces, (which normally involves molecular fracture) and the dissipation of energy in inelastic deformations (which do *not* normally involve molecular fracture).

l) Macroscopic fracture may thus be sub-divided into Regime I, or nucleation, and Regime II or propagation. For specimens containing pre-existing cracks, Regime II will pre-dominate.

How far the fracture parameters of a specimen, such as lifetime under stress and fracture strength, depend on Regime I or Regime II processes will depend on the material's physical properties, the specimen geometry (*i.e.* are there pre-existing cracks) and the conditions of test (rate, temperature, environment).

m) Molecular fracture is probably important both in Regime I and Regime II, even in polymers where ESR signals cannot be generated by tensile testing. The *role* of molecular fracture, and its interaction with other physical mechanisms such as mechanical loss processes, is however significantly different in the two regimes.

4. Radical Formation, Identification and Decay

4.1. Identification of Radical Species

A typical first derivative spectrum from *cis*-polyisoprene tensile tested at 77 K in a nitrogen atmosphere, is shown in Fig. 32 [15, 50]. Identification of the radical species does not follow a prescribed route, which automatically leads to an unequivocable evaluation of the radicals giving rise to the spectrum. There are, however, certain approaches which can be adopted and which lead to a plausible assignment of the spectra. Such assignments are made with due consideration of the most likely sites

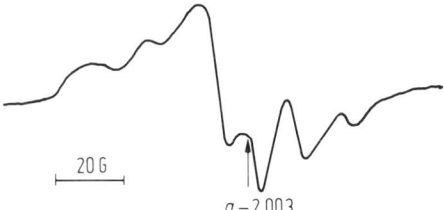

Fig. 32. First derivative ESR spectrum at 123 K from peroxide cross linked *cis*-polyisoprene, tensile tested at 77 K in an oxygen free nitrogen atmosphere

for molecular cleavage and by comparison of the observed spectrum with theoretically predicted curves. The identification of radicals in solid polymers is often made more difficult due to

a) line broadening, resulting in a poorly defined spectrum, and

b) the superimposition of several spectra from different radical species.

It is not within the scope of the present review to give details of the identification procedures. However, a brief indication of the techniques employed will be given. Fuller descriptions are provided in the various texts[31, 32].

The spectrum from *cis*-polyisoprene tensile tested in (nominally) oxygen free nitrogen, shown in Fig. 32, is a six line spectrum with approximately 12 G hyperfine splitting. Polymer radicals are centred around the free spin *g*-value ($g = 2$) and structure in the spectrum is invariably due to nuclear hyperfine structure. The peroxy species is a notable exception and is formed by the reaction $\dot{R} + O_2 \rightarrow R\dot{O}_2$. The spectrum for a peroxy radical arises from *g*-factor anisotropy, and is shown in Fig. 33. The presence of peroxy radicals with their superimposed spectrum, often leads to asymmetry in the overall spectrum and a characteristic line at a *g*-value of 2.033. Such is the situation in Fig. 32. If it is suspected that the observed spectrum is a composite of two or more spectra, one spectrum may be subtracted from the other in an attempt to determine the component species. Spectral subtraction of the peroxy radical spectrum from that shown in Fig. 32 results in the symmetrical 6-line spectrum shown in Fig. 34.

Many of the fundamental spectra obtained from common plastics have been analysed and are quoted in the literature[31, 33, 44, 73]. Analysis in these cases has

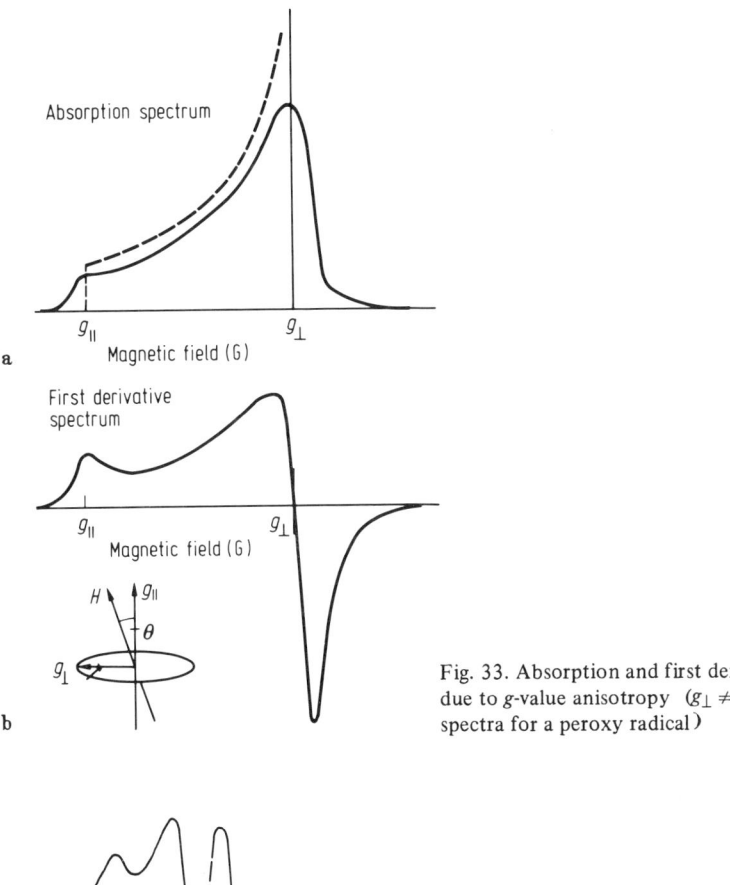

Fig. 33. Absorption and first derivative spectra due to *g*-value anisotropy ($g_\perp \neq g_\parallel$, similar to spectra for a peroxy radical)

Fig. 34. First derivative spectrum from tensile tested *cis*-polyisoprene, after spectral subtraction of the peroxy radical content

normally proceeded by constructing theoretical spectra and comparing them with experimentally obtained spectra. To compute an absorption spectrum it is necessary to assume a line shape, a position function (*i.e.* hyperfine splitting constant), and a relative intensity, for each line in the spectrum. The line shape is taken to be either Gaussian or Lorentzian. It is convenient to handle all calculations in terms of dimensionless parameters, by referring to the half-width of the absorption line $\Delta\omega_{1/2}$. The shape function for the absorption curve is then expressed in terms of the parameter

$$x = \frac{\omega - \omega_0}{\Delta\omega_{1/2}} \tag{25}$$

where ω, ω_0 and $\Delta\omega_{1/2}$ are shown in Fig. 35.

The hyperfine separation of the various lines, a_i, is expressed by $k_i = a_i/\Delta\omega_{1/2}$. The intensity function is simply the nuclear multiplicity for the particular transition involved.

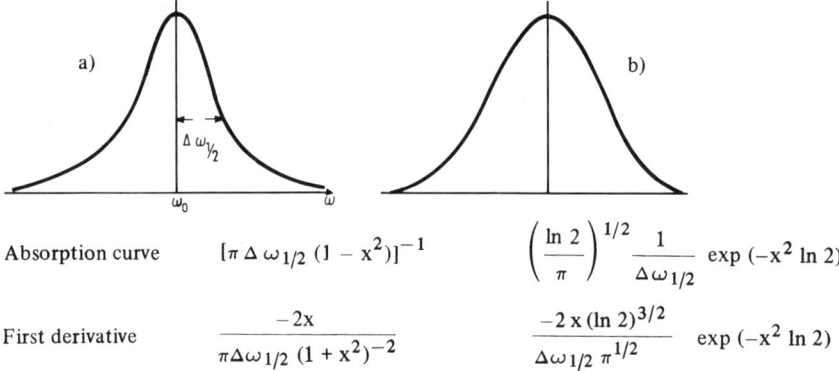

	a)	b)
Absorption curve	$[\pi \Delta\omega_{1/2} (1 - x^2)]^{-1}$	$\left(\frac{\ln 2}{\pi}\right)^{1/2} \frac{1}{\Delta\omega_{1/2}} \exp(-x^2 \ln 2)$
First derivative	$\dfrac{-2x}{\pi\Delta\omega_{1/2} (1 + x^2)^{-2}}$	$\dfrac{-2x(\ln 2)^{3/2}}{\Delta\omega_{1/2} \pi^{1/2}} \exp(-x^2 \ln 2)$

Fig. 35. Characteristic line shape for a) Lorentzian and b) Gaussian absorption curve, together with functions for absorption and first derivative curves

Figure 35 shows the form and shape functions for both the absorption and first derivative curves, for both the Lorentzian and Gaussian assumptions. To generate an individual line, it is necessary only to assume the specific line shape and the half width of the curve. The spectrum arising from the superimposition of several lines is achieved by summing the contributions from all the lines, weighted by the appropriate nuclear multiplicities and assigned appropriate hyperfine separations. The line shape function of the first derivative curve for a set of Lorentzian curves, with the interaction of N_1 protons of hyperfine splitting $a_1 = k_1 \Delta\omega_{1/2}$ and a further N_2 protons of hyperfine splitting $a_2 = k_2 \Delta\omega_{1/2}$ is

$$F'(x) = \sum_{r_1=0}^{N_1} \sum_{r_2=0}^{N_1} {}_{N_1}C_{r_1} \, {}_{N_2}C_{r_2} \left\{ x - \left(\frac{N_1}{2} - r_1\right) k_1 - \left(\frac{N_2}{2} - r_2\right) k_2 \right\} \times$$

$$\left\{ 1 - [x - \left(\frac{N_1}{2} - r_1\right) k_1 - \left(\frac{N_2}{2} - r_2\right) k_2]^2 \right\}^{-2} \tag{26}$$

Such expressions are clearly suitable for computational analysis. Theoretical curves can be generated by randomly adopting hyperfine splittings, nuclear multiplicities etc. Such exercises, although curve fitting operations, can reveal useful information on the possible make-up of an experimental spectrum.

Theoretical considerations can provide values for the hyperfine coupling constants and spin intensities to be used in different situations and the reader is referred to authorative texts on the subject for further guidance[31].

Before seeking values for the hyperfine coupling constant, it is necessary to consider the chemistry of the polymer, the possible locations of the unpaired electron and hence the protons which may give rise to nuclear hyperfine structure.

Identification of the radicals arising from the γ-irradiation of nylon 66 was carried out by Verma and Peterlin[73]. After leaving the specimens for 3 days at 300 K, the first derivative spectrum shown in Fig. 36 was obtained. Three peaks, corresponding to three lines are obvious. A further two clear shoulders and a possible third suggest

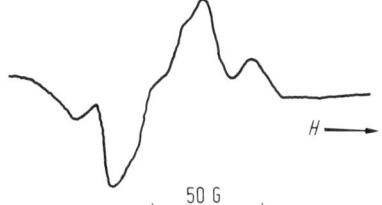

Fig. 36. First derivative ESR spectra of nylon 66 after irradiation at 77 K and then maintaining specimen at 300 K for 3 days (after Ref. [73])

that the spectrum is possibly a six-line spectrum. Analysis of the line intensities showed that the six lines had an intensity ratio of 1:1:2:2:1:1. Consideration of the chemistry of nylon 66 shows that one possible radical species was

$-CO-NH-\dot{C}H-CH_2-$

In this location the electron spin can couple with the nuclear spin of one α-proton and two β-protons. The β-protons are equal, and thus lead to a binomial distribution between the different energy levels, as shown in Fig. 37a with relative line intensities of 1:2:1. The coupling constant for the β-protons had been previously demonstrated to be of the order of 28 G. Interaction with the α-proton causes further splitting of the energy levels. The coupling constant for the latter case was established as varying between 14 G and 26 G, depending on the orientation of the sample to the magnetic field. Assuming a value of 17 G for the splitting constant of the α-proton, Verma and Peterlin obtained a theoretical six line spectrum of relative line intensities 1:1:2:2:1:1, and with the line separations as indicated in Fig. 37b. Hence they assigned the observed spectrum to the radical

$-CO-NH-\dot{C}H-CH_2-$

with the free spin located adjacent to the amide group.

A theoretical analysis of the spectra emanating from UV irradiation of cis-1,4-polyisoprene is provided by Carstensen[74]. Carstensen reasoned that two possible

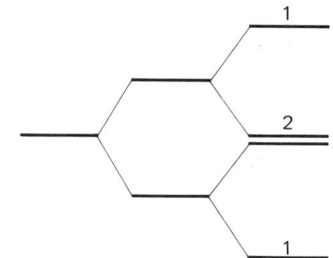

Fig. 37. Theoretical derivation of absorption spectrum from 2 β protons and 1 α proton to give spectrum of relative intensity 1:1:2:2:1:1

radicals could be formed by main chain scission which, with their resonant forms would give

$$\text{(I)} \quad -CH_2-\underset{\underset{CH_3}{|}}{C}=CH-\dot{C}H_2 \leftrightarrow -CH_2-\underset{\underset{CH_3}{|}}{\dot{C}}-CH=CH_2$$

$$\text{(II)} \quad \cdot CH_2-\underset{\underset{CH_3}{|}}{C}=CH-CH_2- \leftrightarrow CH_2=\underset{\underset{CH_3}{|}}{C}-\dot{C}H-CH_2-$$

The number of possible interacting protons for the two different radicals are as follows,

Number of interacting protons

$- C - C = C - \dot{C} - C -$

$ \beta_2 \alpha \beta_1 \alpha \beta_2$

Radical	Number of protons in position		
	α	β_1	β_2
(I)	2	1	5
(II)	3	—	2

Detailed calculations give the spin densities on the α, β_1 and β_2 carbon atoms as 0.622, −0.231 and 0.622 respectively. The hyperfine separation is given by

$$a = Q'\rho \tag{27}$$

where Q' is the splitting parameter and ρ the spin density. The splitting parameter was estimated as 23 gauss, giving hyperfine separations of 14 gauss due to hydrogen atoms attached at the α and β_2 positions, but only 5 gauss for the hydrogen atom at the β_1 position. Carstensen argues that separate lines due to the β_1-proton could not be resolved, and would merely be a source of line broadening. Since the α_1 and β_2 protons are considered equal, each leading to a 14 G splitting, it follows that Radical (I) should lead to an 8 line spectrum and Radical II to a 6-line spectrum. Separately the two spectra would be expected to have line intensities which follow a binomial distribution. However, the observed spectrum would arise from the superimposition of the two spectra. Hence a 14 gauss separation between the lines would be expected, but the line intensities would not follow a binomial distribution. The anticipated observed spectrum would contain 6 or 8 lines of 14 gauss separation.

Figure 38 shows results obtained by Carstensen from UV irradiated cis-1,4-polyisoprene at 77 K. Only 4 lines were clearly defined in the first derivative curve. Figure 38 shows the result of integrating the first derivative curve to obtain

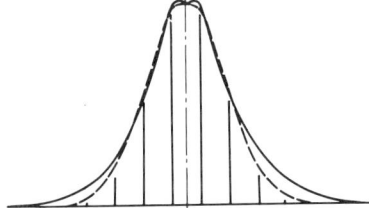

Fig. 38. ESR spectrum of cis-1.4-poly(isoprene) irradiated with UV light at 77 K and measured immediately after irradiation. Full line shows the integrated curve; the dotted line shows the theoretical shape of a spectrum consisting of 8 Gaussian distributed lines with the intensity distribution 1:11:41:75:75:41:11:1

the corresponding absorption curve, and the comparison of the experimentally obtained absorption curve with the theoretically determined 8 Gaussian line spectrum (dotted) with intensity distribution 1:11:41:75:75:41:11:1. Clearly there is close similarity between the experimental and theoretically determined spectrum which strongly suggests that the observed spectrum does arise from the radicals formed by main chain fracture at the weakest bond.

Mead et al.[15, 50] carried out tensile tests on oriented cis-1,4-polyisoprene at 77 K. The first derivative spectrum obtained, after deduction of the peroxy radical, is shown in Fig. 34. In this case, six lines are clearly visible in the experimentally obtained spectrum. The six lines are again equi-separated with 12–14 G splitting, and comparison with the theoretical spectrum shows that it too can be attributed to radicals formed as a result of main chain scission between the α-methylene groups.

4.2. Primary and Secondary Radical Species

Due to the mobility and reactivity of unpaired electrons, it is very common for the free spin to move along the polymer chain or to react with its environment. When this occurs, the initial (primary) radical formed is changed to a more stable secondary radical. Evidence of this is shown by the changes that occur in the ESR spectrum,

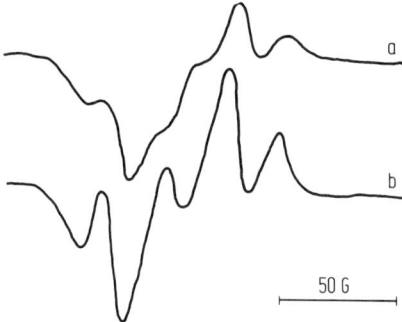

Fig. 39. First derivative ESR spectra of nylon 66 a) irradiated and measured at 77 K b) irradiated at 77 K and measured immediately upon warming to 300 K (after Ref.[73])

with changes in specimen temperature and time. Fig. 39a shows the first derivative spectrum obtained by Verma and Peterlin[73] from γ-irradiated nylon 66, taken at 77 K, while Fig. 39b shows the spectrum obtained after warming the specimen to 300 K. The spectrum obtained at low temperature was assigned to primary radicals arising from main chain rupture

$$\begin{matrix} & H & & & H & H \\ & | & & & | & | \\ -C-N-C\cdot & & \text{and} & & \cdot C-C- \\ \| & | & | & & | & | \\ O & H & H & & H & H \end{matrix}$$

The second spectrum obtained at 300 K was attributed to the secondary radical

$-CO-NH-\dot{C}H-CH_2-$

formed by electron transfer along the chain. The greater stability of the primary radical species at low temperatures is a further reason for conducting experiments at 77 K or lower temperatures.

Table 1 lists the assignments of observed spectra obtained from a range of polymers mechanically degraded by milling at low temperatures. Table 2 lists primary and secondary radicals observed in stressed fibres and filaments.

Reference to Tables 1 and 2 shows that the primary radicals invariably arise from main chain rupture, when radicals are produced by mechanical degradation. When irradiation is employed, primary radicals can also arise from the removal of side groups attached to the main chain. It is evident, therefore, that primary radical formation in mechanical degradation is always attributable to the polymer network being stretched sufficiently to rupture the chains, rather than molecules sliding over each other to strip off side groups.

ESR spectroscopy cannot be used to monitor uniquely the rupture of an individual molecule in a small region. It can only monitor the aggregate of radicals, and hence molecular rupture, over a large region. However, supplementary techniques can be employed to define the location of the molecular rupture site more closely. Becht and Fischer[24] diffused methacrylate monomer into a semi-crystalline polymer, presuming that the monomer would enter only the amorphous regions. Find-

Table 1. Identification of primary and secondary radicals formed by mechanical degradation

Degraded polymer	Sample prepared by milling at °K	ESR spectra taken at °K	Assignment of spectra to Primary radicals	Secondary radicals	Ref.
1. Polyethylene (PE) $(-CH_2-CH_2-)$	77 240 80–100	77 240 77	$-CH_2-H_2\dot{C}$		43, 75, 76) 39) 41)
2. Polypropylene (PP)	80–100 77	77 77	$-CH_2-(CH_3)H\dot{C}$		41) 43)
3. Polyisobutylene (PiB) $[-CH_2-(CH_3)_2C-]$	77 77	77 77	$-CH_2-(CH_3)_2\dot{C}$		77) 39)
4. Polystyrene (PS) $[-CH_2-(C_6H_5)CH-]$	77 77 —	77 77 300	$-CH_2-(C_6H_5)H\dot{C}$	$RO\dot{O}$ $-CH_2-(\dot{C}_6H_4)CH_2$	39) 40, 78) 79)
5. Poly-α-methylstyrene (PαMS) $[-CH_2-(C_6H_5)(CH_3)C-]$	77[a] 77[a]	— 177	$-CH_2-(CH_3)(C_6H_5)\dot{C}$	$-CH_2-(\dot{C}_6H_4)CH_2$	80) 77)
6. Polyvinylacetate (PVAc) $[-CH_2-(OCOCH_3)CH-]$	77 77	77 77	$-CH_2-(OCOCH_3)H\dot{C}$		40) 39)
7. Polymethacrylate (PMA) $[-CH_2-(CO_2CH_3)CH-]$	77 77[a]	77 77	$-CH_2-(COOCH_3)H\dot{C}$		39, 43) 78)
8. Polymethyl methacrylate (PMMA) $[-CH_2-(CO_2CH_3)(CH_3)C-]$	77 77	77 77	$-CH_2-(COOCH_3)(CH_3)\dot{C}$ $-CH_2-(COOCH_3)(CH_3)\dot{C}$ $-CH_2-(COOCH_3)(CH_3)\dot{C}$ to 80–90% $-\dot{C}H-(COOCH_3)(CH_3)CH$ few %	$R_1-\dot{C}H-R_2$	39) 40) 39) 77)
9. Polycarbonate (PC) $[-O-C_6H_4-C(CH_3)_2-C_6H_4-O-CO-]$	77	77	$-O-C_6H_4$ and $-OC-O-H_4\dot{C}_6$		81)

Table 1 (continued)

Degraded polymer	Sample prepared by milling at °K	ESR spectra taken at °K	Assignment of spectra to Primary	Secondary radicals	Ref.
10. Polyamides (PA)					
Polycaprolactam [−CO−(CH$_2$)$_5$−NH−]	77	77	−CH$_2$−H$_2$Ċ 50%		43, 63, 75, 76
	77	113	−CO−H$_2$Ċ and −NH−H$_2$Ċ each 25%	−NH−ĊH−(CH$_2$)$_4$−CO−	43, 82, 76
Poly-α-methyl caprolactam [−CO−CH(CH$_3$)−(CH$_2$)$_4$−NH−]	77	77	−CO−ĊH(CH$_3$)		75)
Poly-ε-methyl caprolactam [−CO−(CH$_2$)$_4$−CH(CH$_3$)−NH−]	77	77	−NH−ĊH(CH$_3$)		75)
11. Thiokol (and other polymers containing sulfide bonds in the main chain)					
(−CH$_2$−S−S−)	77	77	R−Ṡ		39, 83, 84
12. Further references concerning primary radicals formed in mechanical degradation of other polymers.					78, 85, 86, 91

[a] Powder prepared from dilute solution.

Table 2. Free radicals observed in stressed fibres and filaments

Polymer	Primary radical	Secondary radical	Ref.
1. Polyethylene	$-CH_2-\dot{C}H_2$	$R\dot{O}_2$	43, 18)
		$H-CH_2-\dot{C}H-CH_2-$	90)
2. Polystyrene and poly (methyl methacrylate)	No. of radicals at fracture below sensitivity of spectrometer		43, 18)
3. Nylon 66	$-CO-NH-\dot{C}H_2$ and $\dot{C}H_2-CH_2$	$-CO-NH-\dot{C}H-CH_2-$	73, 43, 18)
4. Polydienes			
cis-1,4-polyisoprene	$-CH_2-C(CH_3)=CH-\dot{C}H_2$ and $-CH_2-\dot{C}H=C(CH_3)-\dot{C}H_2$	$R\dot{O}_2$	15, 50, 74)
Polybutadiene	$-CH_2-CH=CH-\dot{C}H_2 \leftarrow \rightarrow -CH_2-\dot{C}H-CH=CH_2$		15, 50, 87, 89)
		$R\dot{O}_2$	
5. Natural silk		$-NH-\dot{C}H-CH_2-$ [a]	88, 18)

[a] At room temperature.

ing that radical formation under stress led to polymerization of the methacrylate monomer, they deduced that molecular rupture and hence the radicals observed by ESR, occurred in the amorphous regions.

4.3. Reactions with Oxygen

Reference to the possible secondary radicals observed in mechanically degraded polymers, shows that the peroxy radical ($R\dot{O}_2$) frequently occurs. Since the polymer molecule contains no oxygen in many cases, the peroxy radical can only arise by reaction of the primary radical with oxygen either dissolved in the specimen or in the test environment. The detection of oxy-radicals ($RO^{\cdot}$) by ESR spectroscopy has been reported in one instance[91] but disclaimed in later papers[92, 93].

Peroxy radicals are formed by the reaction

$$R^{\cdot} + O_2 \rightarrow R\dot{O}_2$$

Structure in the ESR spectrum for peroxy radicals arises from g-tensor anisotropy, and can give rise to different spectra under different situations. Spectra for polypropylene peroxy radicals at four different temperatures are shown in Fig. 40 and are due to Chien and Boss[94]. Which spectrum obtains, depends on the mobility of the molecules containing the peroxy radical species. Increasing mobility leads to an averaging of the g-values and a reduction of the number of lines in the spectrum, until only a broad singlet is observed.

The kinetics of macroradical oxidation have been studied by Bresler and Kazbekov[95], who found that the oxidation rate was essentially constant with time. The

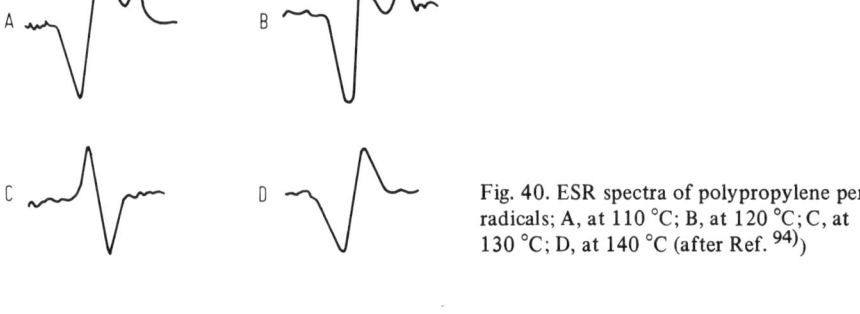

Fig. 40. ESR spectra of polypropylene peroxy radicals; A, at 110 °C; B, at 120 °C; C, at 130 °C; D, at 140 °C (after Ref. [94])

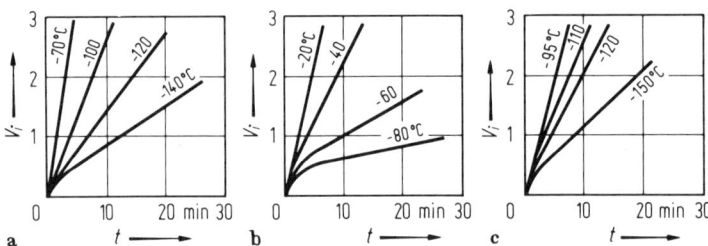

Fig. 41. Kinetics of macroradical oxidation a) poly (vinyl-acetate); b) poly (methyl methacrylate) c) polystyrene (after Ref. [95])

oxidation of polyvinylacetate, poly(methymethacrylate) and polystyrene with time is shown in Fig. 41 and generally follows relationships of the form

$$V_i = \ln \frac{N_\infty}{N_\infty - N} = kt \tag{28}$$

where N_∞ is the final number and N the number of peroxide radicals at time t and k is the rate constant. It appears that oxidation measured by Bresler and Kazbekov occurred only at the surfaces of the specimen, since the above relationship does not hold if there is significant diffusion of oxygen into the polymer. However, crazing occurs in polymers when tensile tested at temperatures within approximately 30 K of the liquefaction temperature of the gaseous environment[97]. Crazing has been observed in polydienes[48, 49] and nylon 66 fibres[98] when they are tensile tested in a nitrogen environment at temperatures between 80–130 K. If the nitrogen contains some oxygen, then the crazes provide easy paths for the oxygen to diffuse through the specimen and react with primary radicals.

The reaction of oxygen with the primary radical species is usually considered a nuisance, since it confuses the spectrum from the primary radical species. Hence the usual procedure is to produce radicals under conditions in which oxygen is excluded as far as possible. Mead, Porter and Reed[50] however, considered that ESR provided a valuable technique for studying oxidation in polybutadiene and polyisoprene. Radicals produced during tensile deformation of oriented samples in nitrogen at a temperature of 83 K were attributed to primary radicals arising from main chain rupture (see Table 2) and peroxy radicals. Mead et al. noted that there were a number of opportunities for oxygen to be absorbed by the polydienes tested, both during the initial preparation stage and during tensile testing. The greater proportion of the peroxy radicals observed were shown to result from oxygen in the tensile test environment reacting with the primary radical species. Figure 42 shows the spectra obtained from tensile tested polybutadiene samples after testing in commercially pure nitrogen and oxygen-free-nitrogen. Figure 43 shows spectra for equivalently tested *cis*-polyisoprene. The spectrum obtained from testing polyisoprene in commercially pure nitrogen is almost wholly attributable to peroxy radicals. The spectrum obtained from polybutadiene after testing in commercially pure nitrogen also contains a high proportion of peroxy radicals but the detail from the primary radical species can be seen in Fig. 42a. The peroxy radicals observed after tensile testing in oxygen-free-nitrogen were attributed to the presence of oxygen in the samples absorbed, or reacting with the rubber, during the initial preparation of the

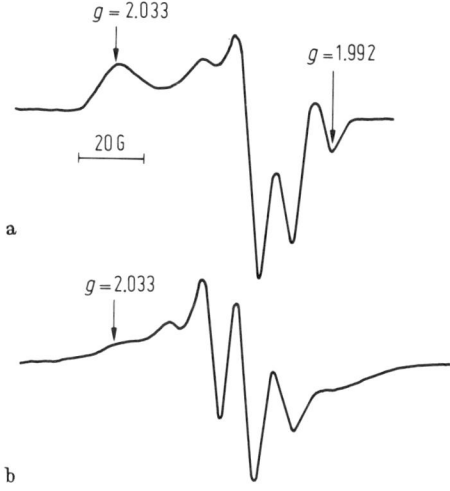

Fig. 42. First derivative spectra ESR spectra recorded at 123 K for polybutadiene following tensile testing a) tensile tested in commercial purity nitrogen; b) tensile tested in oxygen free nitrogen

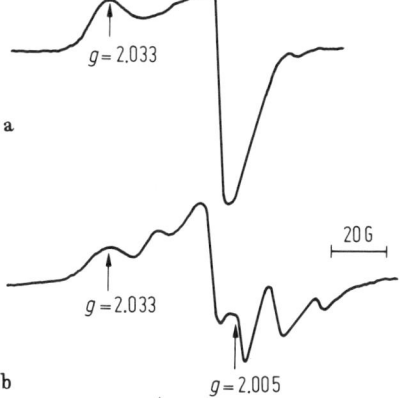

Fig. 43. First derivative ESR spectra from *cis*-polyisoprene following tensile testing at 83 K. a) tensile tested in commercial purity nitrogen; b) tensile tested in oxygen free nitrogen

rubbers. The materials were prepared on an open mill, and oxygen could easily be absorbed at this stage. Although evacuation of the samples before testing was carried out, it made little difference to the peroxy radical content. It was pointed out that it would be difficult to pump off all the absorbed oxygen and only a small residual quantity is necessary for the production of peroxy radicals.

Mead et al.[50] found that the residual peroxy radical content could be almost eliminated by heating the samples to 393 K in oxygen free nitrogen for 15 mins. prior to tensile testing in oxygen free nitrogen at 83 K. They suggested that the heat treatment destroyed hydroperoxides (ROOH) which had previously formed during the initial preparation of the rubbers. Thus the peroxide radicals observed were mainly attributed to reaction of primary radicals with oxygen in the test environment, but some were considered to arise from the hydroperoxides. The mechanism of peroxy radical formation from hydroperoxides was not discussed. Rabek and Rånby[96], in discussing the photoxidation of polymers, point out that the energetically most favourable bond to rupture in the hydroperoxide would result in oxy-radicals (RO·) and hydroxyl radicals (HO·) rather than peroxy radicals. Hence the peroxy radicals observed in the polydienes would have to result from secondary reactions if their origin is indeed due to the presence of hydroperoxides.

Total radical concentrations can be determined by double integration of the first derivative spectrum and comparison with similarly treated specimens of known radical concentration. Diphenylpicrylhydrazyl (DPPH) is commonly used as a reference sample. However, it should be noted that radical concentrations in the reference sample are only quoted to within ± 25% usually, hence accurate determination of the absolute radical concentration in the unknown sample is not possible. Relative radical concentrations of different samples, all of which have been compared to the same reference sample, are not subject to the same error.

Mead et al.[50] also studied the relative stability of the peroxy radical with increasing temperature, by measuring the total radical concentration over a range of temperature and by determining the percentage of the total radical concentration that could be attributed to peroxy radicals. The peroxy radical content was established using spectral deduction techniques. Results obtained from *cis*-polybutadiene samples tensile tested in commercial purity nitrogen are shown in Fig. 44. The high initial peroxy content (~ 90%) reflects the ease with which primary radical species react with molecular oxygen; throughout the specimen in this instance, due to crazing. Peroxy radicals readily form at ~ 80 K in polydienes. Reference to Fig. 44 shows that the peroxy radicals were found to be less stable than the other radicals present, the peroxy radicals decaying rapidly at temperatures 20–30 K below the T_g of the polymer.

Termination reactions listed by Rabek and Rånby[96] are

$$\left. \begin{array}{l} R_1\dot{O}_2 + R_2\dot{O}_2 \\ R_1\dot{O}_2 + R_2^{\cdot} \\ R_1^{\cdot} \phantom{\dot{O}_2} + R_2^{\cdot} \end{array} \right\} \rightarrow \text{inactive products}$$

where the proportion of peroxy radicals is large, it was suggested that the first of these three reactions predominates.

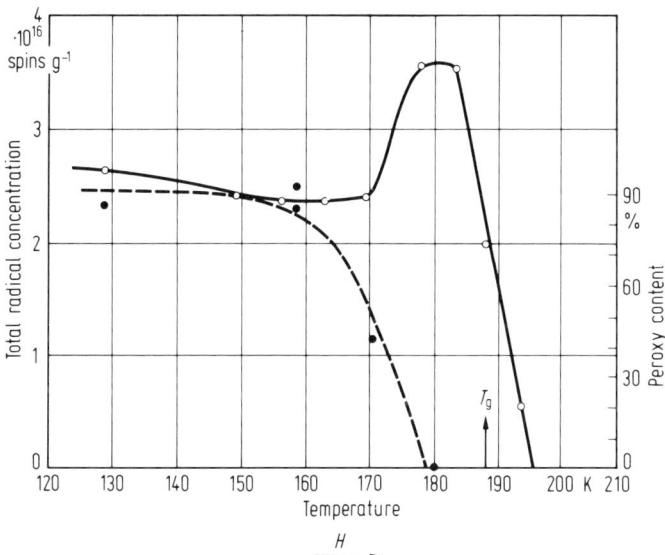

Fig. 44. Variation of total radical concentration and peroxy radical content (dotted) with temperature for *cis*-polybutadiene following tensile testing at 83 K in commercial purity nitrogen

4.4. Radical Decay

Reference has previously been made in Section 2 to the rapid decay of all radical species in the vicinity of the T_g of the polymer. Data for the combined decay rates of all radical species in polybutadiene at various temperatures are shown in Fig. 11. The decay rate at temperatures well below T_g is negligible, but increases rapidly in the range $(T_g - 30)$ K to T_g, where the decay becomes very rapid. The increasing decay rate with temperature in the vicinity of T_g is clearly associated with molecular mobility, particularly main chain rotation. Since the radicals observed arise from main chain scission, the radicals will be near or at chain ends. It is possible that the chain ends will become mobile at temperatures just below the temperature at which main chain rotation freely occurs, and this may account for the observed radical decay at temperatures just below T_g. Alternatively, it may be attributable to the general impreciseness of the glass transition temperature.

Figure 11 shows that the decay rate decreases with the lapse of time. This would naturally follow, if the mechanism of decay is associated with reactions listed at the end of the previous section. After "geographically" easy combinations of free radical species occur, it becomes increasingly difficult for the unpaired electrons to meet and combine. The decay rate consequently declines with time.

Radical decay associated with combination of radical species can result in cross-linking. Such reactions are used commercially to cross-link polyethylene, following irradiation. Cross-linking results in an improvement in the mechanical properties of the material, as is the case in irradiated polyethylene. It might be expected, therefore, that specimens which have been tensile tested under conditions which produce free radicals, might exhibit an improvement in their mechanical properties (modulus, yield stress etc.), after being warmed through the T_g so that cross-linking occurs. Mead[15] was unable to detect any such improvements in *cis*-polybutadiene. However, the number of radicals measured represented only ~ 0.01% of the total number of network chains in the rubber assuming each pair of radicals signified one broken chain. Hence the damage effected on the molecular network was relatively slight and would not be expected to cause significant modifications to the mechanical behaviour.

Studies of the variation in radical concentration with increase in temperature shows the very distinctive rise in radical concentration around T_g, when the radicals are produced by mechanical degradation. This effect is clearly visible in Fig. 44 and is referred to as the anomalous effect or anomalous decay. It is not observed from samples in which the radicals have been produced by irradiation. Figure 45 shows the decay curves obtained by Sohma *et al.*[90, 99] from polypropylene, where the radicals have been produced by mechanical degradation or γ-irradiation. The peak in radical concentration is absent in the case of γ-irradiated samples.

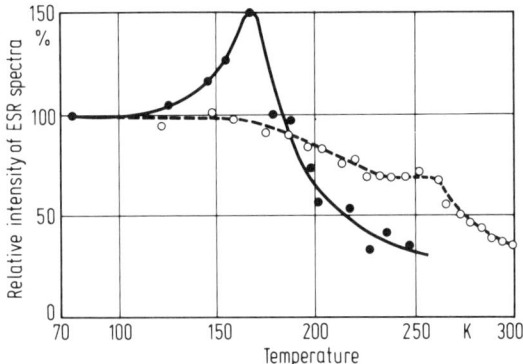

Fig. 45. Decay curves of polypropylene radicals. The black circles and open circles correspond to radicals produced mechanically and by γ-irradiation respectively

Although a complete explanation of the anomalous decay has still to be achieved, several important factors relating to it have been established. In samples where the radicals are produced by milling and grinding, Sakaguchi and Sohma[99] have found that the anomalous decay is connected with electric charges produced at the surfaces of the ground particles by tribo-electricity. Ground samples were found to be negatively charged. Earthing the samples so that the surface charge could escape, resulted in a significant reduction in the peak of the anomalous decay. Whether triboelectricity is a factor contributing to the anomalous decay in tensile tested samples, is open to debate. It is difficult to envisage similary large amounts of friction in the latter case.

Tensile testing results in an elongation of the specimens. This deformation is retained at low temperatures, but retraction of the samples occurs as the T_g is approached. If the retraction is unrestrained, this results in more material being drawn into the active zone of the ESR cavity. An apparent increase in radical concentration can result. Mead et al.[50] found that the anomalous decay observed in tensile tested samples is reduced, but not eliminated, by maintaining the tensile tested samples in the stretched condition as they are warmed through T_g.

Both Mead et al.[50] and Sakaguchi and Sohma[99] agree that oxygen plays an important function in the anomalous decay. Mead showed that tensile testing in oxygen free nitrogen, plus restraining the samples during warming, practically eliminated the anomalous effect in that case. Sakaguchi and Sohma considered that oxygen plays two important roles in the anomalous decay associated with ground, sawn or milled samples. Sohma proposed that the excess charge produced by tribo-electricity led to the formation of carbonions by reaction with the primary radicals

$R\cdot + (-e) \rightarrow R^-$ (no ESR spectra)

The presence of oxygen served to release the excess electrons from the carbon-ion and simultaneously form peroxy radicals.

$R^- + O_2 \rightarrow RO_2^\cdot + (-e)$

This latter reaction would result in an increase in ESR intensity and hence would explain the anomalous decay. Some doubt exists whether this proposed mechanism can explain the anomalous decay experienced in tensile tested samples. Firstly the friction effects to produce the tribo-electricity charge are not so clearly present. Secondly, Mead et al.[50] have shown that the peroxy radical content decreases as the rise in the total radical concentration occurs. Clearly the anomalous decay cannot be explained by peroxy radical formation in the latter case.

Since the radicals decay rapidly in the region of the T_g of the material, it is difficult to determine accurately the radical concentration in this region. Radical concentrations measured around T_g are a function of the time at which recordings were made after the radicals were formed, and are subject to the possibility of secondary radicals being formed. If it is assumed that the initial radicals decay in quantity without modification, techniques exist[20] to extrapolate the radical concentration versus time data to determine the initial radical concentration. One technique used is to extrapolate curves as shown in Fig. 11. However, care must be taken to ensure that zero time on the plot truly corresponds to the onset of radical formation.

5. Conclusion

In this review we have attempted to set the subject of molecular fracture in the context of the various theories applying to macroscopic or engineering fracture. Two regimes of fracture have been identified, namely a thermally activated Regime I dominated by bond fracture processes and a mechanical Regime II which also involves molecular fracture *via* the surface energy term but in which mechanical loss processes control the resistance to fracture and its rate/temperature dependence.

Experimental studies of molecular fracture cannot be used directly to predict macroscopic fracture properties, for various reasons. Firstly, in Regime II, resistance to fracture is controlled largely by flow processes which do not involve molecular breakage. Secondly, even in Regime I, where mechanical losses are negligible, there are processes intermediate between molecular fracture events and macroscopic failure. Such processes include the redistribution of load between load-bearing molecules, microvoid formation and growth and crazing.

In spite of the difficulty of establishing direct correlations between macroscopic fracture phenomena and microscopical molecular fracture phenomena, the large amount of work carried out in recent years on the application of ESR and other techniques to molecular events in stressed polymers has brought us much closer to success. In particular it can now be stated in broad terms but with some confidence just how far the breakage of molecules within a stressed polymer contributes to macroscopic deformation and failure in any particular case. Work will no doubt continue with a view to reconciling even more closely the microscopical and macroscopic descriptions of fracture phenomena.

6. References

[1] Cottrell, A. H.: Proc. Roy. Soc. (Lond.) A *282*, 2 (1964)
[2] Andrews, E. H.: Fracture in polymers. London: Oliver & Boyd Ltd. 1968, p. 113
[3] Griffith, A. A.: Phil. Trans. Roy. Soc. (Lond.). A *221*, 163 (1921)
[4] Andrews, E. H.: J. Materials Sci. *9*, 887 (1974)
[5] Orowan, E.: Proc. Symposium on Fatigue and Fracture of Metals New York: John Wiley & Sons Inc. 1950, p. 139
[6] Benbow, J. J., Roesler, F. C.: Proc. Phys. Soc. (Lond.), B, *70*, 201 (1957)
[7] Irwin, G. R.: J. Appl. Mech. *24*, 361 (1957)
[8] Andrews, E. H., Fukahori, Y.: J. Materials Sci. *12*, 1307 (1977)
[9] Lake, G. J., Thomas, A. G.: Proc. Roy. Soc. (Lond.) A *300*, 108 (1967)
[10] Lake, G. J., Lindley, P. B.: J. Appl. Polym. Sci. *9*, 1233 (1965)
[11] King, N. E.: Failure of Metal-to-Epoxy Resin Bonds, Ph. D. Thesis, Univ. of Lond. (1976)
[12] Braden, M., Gent, A. N.: J. Appl. Polym. Sci. *3*, 90 (1960)
[13] Andrews, E. H., Bevan, L.: Polymer *13*, 337 (1972)
[14] Zhurkov, S. N., Kuksenko, V. S., Slutsker, A. I.: Proc. 2nd. Int. Conf. on Fracture, Brighton, 1969 London: Chapman & Hall Ltd., p. 531
[15] Mead, W. T.: Molecular Fracture in Mechanically Deformed Polymers, Ph. D. Thesis, Univ. of London (1975)
[16] Zhurkov, S. N., Vettegren, V. I., Korsukov, V. E., Novak, I. I.: Proc. 2nd Int. Conf. on Fracture Brighton, 1969 London: Chapman & Hall Ltd., p. 545
[17] Knauss, W. G.: Ph. D. Thesis, California Inst. of Technol. Pasadena (1963) and Appl. Mechanics Review (Jan. 1973)
[18] Zhurkov, S. N., Tomashevsky, E. E.: Physical basis of yield and fracture. London: Institute of Physics 1966, p. 200
[19] Dobrodumov, A. V., El'yashevich, A. M.: Eng. Transl. Sov. Phys. Solid State *15*, No. 6, 1259 (1973)
[20] Kausch, H. H., De Vries, K. L.: Int. J. of Fracture *11*, 727 (1975)
[21] Kausch, H. H., Becht, J.: Rheologica Acta *9*, 137 (1970)
[22] Taylor, N. W.: J. Appl. Phys. *18*, 943 (1947)
[23] Casale, A., Porter, R. S., Johnson, J. F.: Rubber Chem. and Tech. *44*, 534 (1971)

24) Becht, J., Fischer, H.: Kolloid-Z. u. Z. Polymere *229*, 167 (1969)
25) Pazonyi, T., Tudos, F., Dimitrov, M.: Angew. Makrom. Chemie *10*, 75 (1970)
26) Salloum, R. J., Eckert, R. E.: J. Appl. Polym. Sci. *17*, 509 (1973)
27) Zhurkov, S. N., Korsukov, V. E.: J. Polym. Sci., Polym. Phys. Ed. *12*, 385 (1974)
28) Zhurkov, S. N., Zakrevskii, V. A., Korsukov, V. E., Kuksenko, V. S.: Engl. Trans. in Soviet Phys. Solid State *13*, 1680 (1972)
29) Bernstein, H. I.: Spectrochim. Acta *18*, 161 (1962)
30) Guinier, A., Fournet, A.: Small angle scattering of X-rays. New York: John Wiley & Sons 1955
31) Ayscough, P.: Electron spin resonance in chemistry. London: Methuen 1967
32) Poole, C. P.: Electron spin resonance. New York: John Wiley and Sons, 1967
33) Campbell, D.: Macrom. Rev. *4*, 91 (1970)
34) Schneider, E. E.: Disc. Faraday Soc. *19*, 158 (1955); J. Chem. Phys. *23*, 978 (1955)
35) Abraham, R. J., Melville, H. W., Ovenall, D. W., Whiffen, D. H.: Trans. Faraday Soc. *54*, 1133 (1958)
36) Charlesby, A., Thomas, D. K.: Proc. Roy. Soc. (Lond.) A, *269*, 104 (1962)
37) Smaller, B., Matheson, M. S.: J. Chem. Phys. *28*, 1169 (1958)
38) Libby, D., Ormerod, M. G., Charlesby, A.: Polymer *1*, 212 (1960)
39) Bresler, S. R., Kazbekov, E. N., Saminskii, E. M.: Polymer Sci. (USSR) *1*, 540 (1959)
40) Zhurkov, S. N., Tomashevskii, E. E., Zakrevskii, V. A.: Engl. Trans. in Soviet Phys. Solid State *3*, 2074 (1962)
41) Radciq, V. A., Butyagin, P. Yu.: Polymer Sci. (USSR) *A9*, 2883 (1967)
42) Verma, G. S. P., Peterlin, A.: Kolloid Z. u. Z. Polymere *236*, 111 (1970)
43) De Vries K. L., Roylance, D. K., Williams, M. L.: Rept. UTEC DO 68-056, College of Eng., Univ. of Utah, Salt Lake City (1968)
44) Kausch, H. H.: J. Macrom. Sci., Revs. Macromol. Chem. C *4*, 243 (1970)
45) De Vries, K. L., Roylance, D. K., Williams, M. L.: J. Polym. Sci. A1, *8*, 237 (1970)
46) Backman, D. N., De Vries, K. L.: J. Polym. Sci. A1, *7*, 2125 (1969)
47) De Vries, K. L., Simonson, E. R., Williams, M. L.: J. Macrom. Sci., Phys. B *4*, 671 (1970)
48) Mead, W. T., Reed, P. E.: Polym. Eng. Sci. *14*, 22 (1974)
49) Mead, W. T., Porter, R. S., Reed, P. E.: J. Materials Sci. *13* (1978)
50) Mead, W. T., Porter, R. S., Reed, P. E.: Macromolecules (1978)
51) Crist, B.: Private Communication
52) Bates, R. F., Jackson, J. B.: Polymer systems-deformation and flow, R. E. Whetton and R. M. Wharlow (eds.) London: MacMillan 1968, p. 153
53) De Vries, K. L., Lloyd, B. A., Williams, M. L.: J. Appl. Phys. *42*, 4644 (1971)
54) Mead, W. T., Reed, P. E.: (to be published)
55) Natarajan, R., Reed, P. E.: J. Polym. Sci., A2, *10*, 585 (1972)
56) Andrews, E. H.: The physics of glassy polymers. R. N. Haward (ed.). London: Appl. Sci. Publishers Ltd. 1973, p. 394
57) Lloyd, B. A., De Vries, K. L., Williams, M. L.: J. Polym. Sci. A2, *10*, 1415 (1972)
58) Park, J. B., Statton, W. O., De Vries, K. L.: Report of the Univ. of Utah. Salt Lake City, UTEC ME 71-122 (1971)
59) Zaks, B. Yu., Lebedinskaya, L., Chalidze, V. N.: Vysokomol. Soyed. A *12*, 2669 (1970)
60) Idem: Engl. Trans. in Polym. Sci. (USSR) *12*, 3025 (1971)
61) Chiang, T. C., Sibilia, J. P.: J. Polym. Sci. *10*, 2249 (1972)
62) Campbell, D., Peterlin, A.: Polym. Letters *6*, 481 (1968)
63) Roylance, D. K., De Vries, K. L., Williams, M. L.: Proc. 2nd. Int. Conf. on Fracture Brighton 1969 London: Chapman & Hall Ltd., p. 551
64) Hassell, W. H.: MS Thesis, Dept. Mech. Eng., Univ. of Utah, Salt Lake City (June 1973)
65) Andrews, E. H.: Angew. Chem. *86*, 151 (1974)
66) Thomas, A. G.: J. Polym. Sci. C, Sympos *48*, 145 (1974)
67) Williams, M. L., De Vries, K. L.: Proc. 5th Int. Congress on Rheology, Ed. S. Onogi, Vol. 3 Tokyo: Univ. of Tokyo Press 1970, p. 139
68) Reeve, B.: Morphology and Tensile Properties of Polychloroprene, Ph. D. Thesis, Univ. of Lond. (1972)

69) Johnson, U., Klinkenberg, D.: Kolloid Z. u. Z. Polymere *251*, 843 (1973)
70) Zhurkov, S. N.: Private communication, and J. Polym. Sci. Polym. Phys. Ed. *12*, 385 (1974)
71) Brown, N., Imai, Y.: J. Materials Sci. *11*, 425 (1976) and references therein
72) Andrews, E. H., Levy, G. M., Willis, J.: J. Materials Sci. *8*, 1000 (1973)
73) Verma, G. S. P., Peterlin, A.: J. Macromol. Sci. Phys. *34*, 589 (1970)
74) Carstensen, P.: Die Makromol. Chemie *135*, 219 (1970)
75) Zakrevskii, V. A., Tomashevskii, E. E., Baptizmanskii, V. V.: Engl. Trans. Soviet Phys. Solid State *9*, 1118 (1967)
76) Zakrevskii, V. A., Baptizmanskii, V. V., Tomashevskii, E. E.: Fiz. Tverdogo Tela *10*, 1699 (1968)
77) Butyagin, P. Yu., Kolbanev, I. V., Radciq, V. A.: Engl. Trans. Soviet Phys. Solid State *5*, 1642 (1964)
78) Dubinskaya, A. M., Butyagin, P. Yu.: Vysokomolekul. Soedin *B9*, 525 (1967)
79) Bueche, F.: J. Appl. Phys. *26*, 1133 (1955)
80) Dubinskaya, A. M., Butyagin, P. Yu.: Vysokomolekul. Soedin. *A10*, 240 (1968)
81) Zakrevskii, V. A., Tomashevskii, E. E.: Vysokomolekul. Soedin. *B10*, 193 (1968)
82) Roylance, D. K.: Ph. D. Thesis, Dept. Mech. Eng., Univ. Utah, Salt Lake City (1968)
83) Zakrevskii, V. A., Tomaskevskii, E. E.: Engl. Trans. Polym. Sci. (USSR) *8*, 1424 (1966)
84) Zhurkov, S. N., Zakrevskii, V. A., Tomashevskii, E. E.: Engl. Trans. Soviet Phys. Solid State *6*, 1508 (1964)
85) Dubinskaya, A. M., Butyagin, P. Yu., Odintsova, R. R., Bolin, A. A.: Vysokomolekul. Soedin. *A10*, 410 (1968)
86) Urbanski, T.: Nature *216*, 577 (1967)
87) Carstensen, P.: Die Makromol. Chemie *142*, 131 (1971)
88) Samoilov, G. G., Tomashevskii, E. E.: Engl. Trans. Soviet Phys. Solid State *10*, 866 (1968)
89) Carstensen, P.: ESR Applications to Polymer Research (Proc. 22nd Nobel Symposium). P. Kinell, B. Rånby and V. Runnstrom-Reis (eds.). New York: John Wiley & Sons, 1973, p. 159
90) Sohma, J., Karvashima, T., Shimada, S., Kashiwabara, H., Sakaguchi, M.: ESR Applications to Polymer Research (Proc. 22nd Nobel Symposium). P. Kinell, B. Rånby and V. Runnstrom-Reis (eds.). New York: John Wiley & Sons 1973 p. 225
91) Piette, L. H., Landgraf, W. C.: J. Chem. Phys. *32*, 1107 (1960)
92) Symons, M. C. R.: Adv. Phys. Organ. Chem. *3*, 307 (1963)
93) Ingold, K. U., Morton, J. R.: J. Am. Chem. Soc. *86*, 3400 (1964)
94) Chien, J. C. W., Boss, C. R. J.: J. Am. Chem. Soc. *89*, 571 (1967)
95) Bresler, S. E., Kazbekov, E. N.: Fortschr. Hochpolym. Forsch. *3*, 688 (1964)
96) Rabek, J. F., Rånby, B.: ESR Applications, to Polymer Research, (Proc. 22nd Nobel Symposium). P. Kinell, B. Rånby and V. Runnstrom-Reis (eds.). New York: John Wiley & Sons 1973, p. 201
97) Brown, N., Imai, Y.: J. Polym. Sci., B, *13*, 511 (1975)
98) Chern, I. S.: private communication
99) Sakaguchi, M., Sohma, J.: Polymer J. *7*, 490 (1975)

Received July 25, 1977

J. D. Ferry (editor)

Applications of Linear Fracture Mechanics

J. Gordon Williams
Department of Mechanical Engineering, Imperial College of Science and Technology, London SW7 2BX, Great Britain

The development of linear elastic fracture mechanics is given with a special emphasis on its application to the testing of polymers. The modelling of crazes and plastic zones is discussed and then developed to describe time-dependent crack and craze growth, including crack stability phenomena.
These results are then applied to particular problems, such as environmental stress cracking, fatigue and impact testing.

Table of Contents

List of Symbols 69
1. Introduction 71
2. Linear Elastic Analysis 71
   2.1. Energy Release Rate, G 71
   2.2. Contour Integrals 74
   2.3. Stresses Around a Crack Tip 75
   2.4. Relationship Between G and K 78
   2.5. Some Solutions for K_I 79
3. Plastic Zones and Crazes 84
   3.1. Small Scale Yielding 84
   3.2. Line Zones and Crazes 85
   3.3. Thickness Effects 89
4. Time Dependence 90
   4.1. Incubation Times 92
   4.2. Slow Crack Growth 93
   4.3. Craze Growth 96
   4.4. Temperature Effects 97
   4.5. Crack Stability 97
5. Environmental Effects 100
   5.1. Crack Growth 101
   5.2. Craze Growth 104

6. Fatigue 111
7. Impact Testing 113
8. Closure 118
9. References 119

List of Symbols

A	Craze pore area
a	Crack length
B	Plate thickness
B_c	Crack width (for grooved specimens)
b	Craze thickness; position of point load
C	Compliance (Δ/P); constant for craze growth; creep compliance function
c	Specific heat
D	Depth of specimen
d_0	Craze parameter
E	Young's modulus
e	Strain
G	Strain energy release rate
G_{IC}	Energy per unit area of crack in mode I
H	Activation energy
K	Stress intensity factor (SIF)
K_I	SIF in mode I
K_{II}	SIF in mode II
K_{IC}	SIF in mode I at fracture
K_{c1}	Plane strain K_{IC}
K_{c2}	Plane stress K_{IC}
k	Thermal conductivity
L	Half span
l_0	Craze parameter
m	Constant; mass of specimen
N	Number of cycles
n	Constant; $(d \ln e)/(d \ln t)$ in creep
P	Load
R	Stress rate; gas constant
r	Coordinate
r_p	Plastic zone size or craze length
r_c	$\pi/8\, K_{IC}^2/\sigma_c^2$
S	Surface length
T	Temperature – absolute degrees
t	Time
U	Energy
u	Displacement or velocity in x direction
V	Velocity
v	Displacement or velocity in y direction
W	Strain energy density
X	$a - \xi$
x	Coordinate or a/D
Y	Finite width correction factor
y	Coordinate
α	Constant; viscoelastic transition
β	Viscoelastic transition
γ	Viscoelastic transition; surface work
Δ	Deflection
δ	Displacement in craze
δ^*	Displacement in craze at crack tip
δ_c^*	Critical value of δ^* at fracture
κ	$3 - 4\nu$ for plane strain
	$3 - \nu/1 + \nu$ for plane stress
λ	Constant

μ	Shear modulus; viscosity
ν	Poisson's ratio
ξ	Working variable – length
ρ	Density
σ	Stress
σ_c	Craze or cohesive stress
σ_y	Yield stress
τ	Working variable – time
ϕ	Stress function; axisymmetric parameter; calibration function

1. Introduction

Linear elastic fracture mechanics (LEFM) describes the behaviour of sharp cracks in linear, perfectly elastic materials. Since polymers are neither linear nor elastic, the utility of the theory may, at first sight, seem doubtful. In fact, the deviations from the theoretical assumptions are such that quite minor modifications to the analysis produce a precise description of crack growth in polymers within the framework of the conventional theory. The considerable resources of the subject may thus be utilised in that testing experience on other materials may be employed, together with the available analytical work.

The value in performing this analysis is that it provides a logical framework within which to describe the fracture behaviour of materials. The parameters used to define toughness, K_{IC} and G_{IC}, are true material properties which may be determined independently of test conditions. The advantage of this over the conventional, *ad hoc*, methods of evaluation currently employed for polymers is considerable, both in material evaluation and the analysis of service failures.

This review will take the form of a fairly complete development of the linear theory from first principles. The necessary special results used for polymers will be incorporated in the analysis and the examples drawn from the polymer literature. There is now a substantial body of work available on the fracture of polymers and this will be used where appropriate. There is no attempt here, however, to give a comprehensive review of the literature since there are several available (*e.g.* [1]). References will be given to those papers which will amplify the necessarily abbreviated treatment given here.

2. Linear Elastic Analysis

The material is assumed to obey Hooke's law so that the stress is proportional to the infinitesimal strains. For any loaded body, therefore, a linear load-deflection relationship will result.

2.1. Energy Release Rate, G

The elements of the theory are probably most easily understood from considering the problem in terms of energy. A simple energy balance argument may be applied to a cracked body of uniform thickness, B, subjected to a generalised load, P, as shown in Fig. 1(a). The linear load-deflection diagram is shown in Fig. 1(b), line (i), so that the stored elastic energy is given by the area under the curve[2]:

$$U_1 = \frac{1}{2} P \Delta \qquad (1)$$

Suppose now that the crack of length a grows by a small amount δa such that the stiffness of the body changes. This will result in a change in load of δP and in

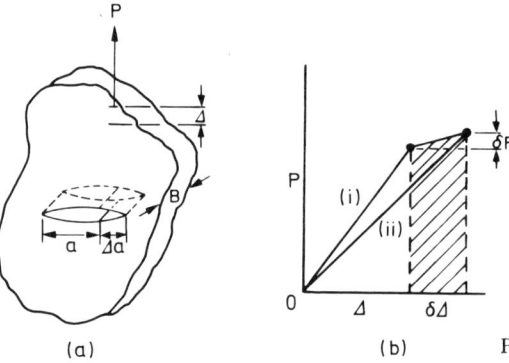

Fig. 1. Loading a general cracked body

deflection of $\delta\Delta$ and, since the material is linearly elastic, unloading would give the line (ii) shown. Thus, the new stored energy is given by:

$$U_2 = \frac{1}{2}(P + \delta P)(\Delta + \delta\Delta) \qquad (2)$$

In addition, there is external work performed [which is shown as the shaded area in Fig. 1(b)], given by:

$$U_3 = \left(P + \frac{\delta P}{2}\right)\delta\Delta \qquad (3)$$

The *change* in energy of the whole system when the crack grows is thus given by:

$$\delta U = U_2 - U_1 - U_3$$

which is the wedge shaped area between lines (i) and (ii) in the diagram. Substituting from Eqs. (1), (2) and (3), we have:

$$\delta U = \frac{1}{2}(\Delta \delta P - P\delta\Delta)$$

If we now define the energy release rate as the energy release per unit area of crack growth, then:

$$G = \frac{\delta U}{B\delta a} = \frac{1}{2B}\left(\Delta\frac{\delta P}{\delta a} - P\frac{\delta\Delta}{\delta a}\right) \qquad (4)$$

If it is now proposed that the energy necessary to grow the crack is G_c per unit area, then fracture can occur when:

$$G \geqslant -G_c$$

Clearly, a crack can initiate when $G = G_c$ and will be unstable if $dG/da < 0$ thereafter. This is the Griffith fracture condition[3] on which all fracture mechanics is based, although Griffith used the surface energy of the crack γ so that:

Applications of Linear Fracture Mechanics

$G_c = 2\gamma$

The fracture initiation criterion can therefore be written as:

$$G_c = -\frac{1}{2B}\left(\Delta \frac{\delta P}{\delta a} - P \frac{\delta \Delta}{\delta a}\right) \tag{5}$$

Some useful results may be obtained by considering the compliance of the cracked body, C, which can be measured or calculated. Now:

$$C = \Delta/P \tag{6}$$

so that:

$$\delta \Delta = P\, \delta C + C\, \delta P$$

and substituting in Eq. (5), we have two results:

$$G_c = \frac{P^2}{2B}\frac{dC}{da} = \frac{\Delta^2}{2B}\frac{1}{C^2}\frac{dC}{da} \tag{7}$$

(The operator d/da is now used since C is a geometric factor in which only a varies.)

In practice, if C is determined as a function of a, then dC/da can be found. If P or Δ at fracture is determined for a given crack length, then G_c can be found from

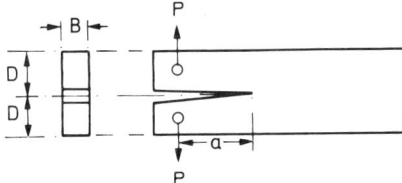

Fig. 2. Cantilever specimen

Eq. (7). The simplest example is the double cantilever beam shown in Fig. 2 for which the deflection may be calculated by simple beam theory:

$$\Delta = 2\left(\frac{Pa^3}{3EI}\right) = \frac{8Pa^3}{EBD^3}$$

so that:

$$C = \frac{\Delta}{P} = \frac{8a^3}{EBD^3}, \quad \frac{dC}{da} = \frac{24a^2}{EBD^3}$$

and:

$$G_c = P^2 \frac{12 a^2}{E B^2 D^3}$$

It should be noted that, in all cases, $C \propto E^{-1}$ so that all the material properties may be collected in one group, so that:

$$E G_c = P^2 \frac{12 a^2}{B^2 D^3} \tag{8}$$

A further general result of some importance may be deduced from Eq. (7) in that, if energy rather than load is determined, then:

$$U_1 = \frac{1}{2} P \Delta = \frac{1}{2} P^2 C$$

so that:

$$G_c = \frac{U_1}{B} \left(\frac{1}{C} \frac{dC}{da} \right) \tag{9}$$

As before, the parameter $C^{-1} dC/da$ may be determined by calibration and G_c found if U_1 at fracture is known (see Section 7).

2.2. Contour Integrals

If the loading applied to the cracked body is distributed as stresses over the boundary instead of being a concentrated load, then the various work terms must be evaluated as integrals taken over the boundary. Thus, if we consider a body as shown in Fig. 3 with normal and shear surface tractions σ_n and σ_s giving dis-

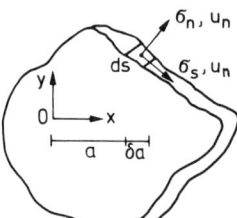

Fig. 3. Distributed loads on a cracked body

placements u_n and u_s over an element of length dS, then the stored energy is given by:

$$U_1 = \frac{B}{2} \int_S (\sigma_n u_n + \sigma_s u_s) \, dS \tag{10}$$

where the integral is taken over the whole contour of the boundary S. The expression for G in Eq. (4) may be rewritten as:

$$G = \frac{1}{2\,\delta a} \int_S (u_n\,\delta\sigma_n + u_s\,\delta\sigma_s) - (\sigma_n\,\delta u_n + \sigma_s\,\delta u_s)\,dS \tag{11}$$

Now the total stored energy, U_1, may also be deduced by integrating the strain energy density function, W, over the whole volume of the body, so that:

$$U_1 = B \int_V W\,dx\,dy$$

where x, y are a cartesian coordinate system, as shown in Fig. 3. When the crack grows δa, the x coordinate of each element in the body will change by δa so that the energy change produced by crack growth may be deduced by replacing δx by δa and we have:

$$\frac{\delta U_1}{\delta a} = B \int_S W\,dy$$

and, if we differentiate Eq. (10), we may write:

$$\frac{\delta U_1}{\delta a} = \frac{B}{2\,\delta a} \int_S (\sigma_n\,\delta u_n + u_n\,\delta\sigma_n) + (\sigma_s\,\delta u_s + u_s\,\delta\sigma_s)\,dS$$

so that:

$$\int_S W\,dy = \frac{1}{2\,\delta a} \int_S (\sigma_n\,\delta u_n + u_n\,\delta\sigma_n) + (\sigma_s\,\delta u_s + u_s\,\delta\sigma_s)\,dS \tag{12}$$

Combining Eqs. (11) and (12) and noting $\delta x = \delta a$, we have the result:

$$G = \int_S W\,dy - \left(\sigma_n\,\frac{du_n}{dx} + \sigma_s\,\frac{du_s}{dx}\right)dS \tag{13}$$

This is the contour integral described by Rice[4] which is usually denoted by J for non-linear elastic materials and becomes G for the linear case given here. The result is important since, if a stress analysis for a cracked body is performed in which the stress and displacement fields are found, then G may be determined.

2.3. Stresses Around a Crack Tip

The stresses around a crack tip are most easily computed[5] from the usual stress function, ϕ, which must satisfy the biharmonic equation:

$$\nabla^4 \phi = 0$$

and the stresses are given by:

$$\sigma_r = \frac{1}{r}\frac{\partial \phi}{\partial r} + \frac{1}{r^2}\frac{\partial^2 \phi}{\partial \theta^2}, \quad \sigma_\theta = \frac{\partial^2 \phi}{\partial r^2}, \quad \sigma_{r\theta} = -\frac{\partial}{\partial r}\left(\frac{1}{r}\frac{\partial \phi}{\partial \theta}\right)$$

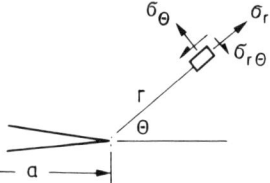

Fig. 4. Stresses around a crack tip

where the stresses are defined in cylindrical polar coordinates with the origin at the crack tip, as shown in Fig. 4. A stress function of the form:

$$\phi = r^n \Theta(\theta)$$

is appropriate here, where $\Theta(\theta)$ may be determined from $\nabla^4 \phi = 0$ as:

$$\Theta(\theta) = [b \sin n\theta + b' \sin(n-2)\theta] + [c \cos n\theta + c' \cos(n-2)\theta]$$

where b, b', c, c' and n are constants.

If the crack faces are taken as stress free, then we have the boundary conditions:

$$\sigma_{r\theta} = \sigma_\theta = 0 \quad \text{at} \quad \theta = \pm \pi$$

so that:

$$\Theta(\theta) = c\left[\cos n\theta - \frac{n}{n-2}\cos(n-2)\theta\right] + b[\sin n\theta - \sin(n-2)\theta] \tag{14}$$

with n taking eigenvalues in the series:

$$n = 0, \pm\frac{1}{2}, \pm 1, \pm\frac{3}{2}, \pm 2, \ldots$$

If the restriction that the displacements shall remain finite at the crack tip ($r = 0$) is imposed, then $n > 1$ so that:

$$n = \frac{3}{2}, 2, \frac{5}{2}, \ldots$$

For example, the hoop stress may now be written as:

$$\sigma_\theta = n(n-1)r^{n-2}\Theta(\theta)$$

Applications of Linear Fracture Mechanics

and we have the series:

$$\sigma_\theta = \frac{3}{4} \frac{1}{\sqrt{r}} \Theta_{n=3/2} + 2 \Theta_{n=2} + \frac{15}{4} \sqrt{r} \, \Theta_{n=5/2} + \ldots$$

If we confine our attention to the region very near to the crack tip, then clearly the first term dominates since it is singular and for this region we have:

$$\sigma_\theta = \frac{3c}{4\sqrt{r}} \left(\cos \frac{3\theta}{2} + 3 \cos \frac{\theta}{2} \right) + \frac{3b}{4\sqrt{r}} \left(\sin \frac{3\theta}{2} + \sin \frac{\theta}{2} \right)$$

The first term is symmetric in θ, while the second is skew symmetric and they represent the opening Mode I and sliding Mode II of crack loading, respectively. Mode I is essentially that for loading normal to the crack, while Mode II is that for shear loading. Since Mode I is by far the most common, we shall confine our attention to the first term which is usually expressed as:

$$\sigma_\theta = \frac{K_I}{\sqrt{2\pi r}} \frac{1}{2} \cos \frac{\theta}{2} (1 + \cos \theta)$$

where K_I is the mode I stress intensity factor and shows that the stress distribution around the crack tip is identical in form for *all* loadings but that its intensity is governed by the constant K_I which is determined by those loads and the geometry. Since $\sigma_\theta \to \infty$ as $r \to 0$, stress does not make a reasonable local fracture criterion. However, the product $\sigma_\theta \sqrt{r}$ remains finite at the crack tip and it may be postulated that $K_I = K_{IC}$, a critical value, could be taken as a fracture criterion.

The full stress and displacement field around the crack tip is given by:

$$\sigma_\theta = \frac{K_I}{\sqrt{2\pi r}} \frac{1}{2} \cos \frac{\theta}{2} (1 + \cos \theta)$$

$$\sigma_r = \frac{K_I}{\sqrt{2\pi r}} \frac{1}{2} \cos \frac{\theta}{2} (3 - \cos \theta)$$

$$\sigma_{r\theta} = \frac{K_I}{\sqrt{2\pi r}} \frac{1}{2} \cos \frac{\theta}{2} \sin \theta \qquad (15)$$

and:

$$u_r = \frac{\sqrt{2\pi r} \, K_I}{8\pi \mu} \left((2\kappa - 1) \cos \frac{\theta}{2} - \cos \frac{3\theta}{2} \right)$$

$$u_\theta = \frac{\sqrt{2\pi r} \, K_I}{8\pi \mu} \left(\sin \frac{3\theta}{2} - (2\kappa + 1) \sin \frac{\theta}{2} \right)$$

where μ is the shear modulus ($E/2(1 + \nu)$), and κ is $3 - 4\nu$ for plane strain and $3 - \nu/1 + \nu$ for plane stress.

There are many texts available (e.g.[6]) which give expressions for K_I for various loads and geometries so that the criterion $K_I = K_{IC}$ may be used in conjunction with these. Some clarity is obtained, however, by relating K_I to G_I since the latter has a more physical basis.

2.4. Relationship Between G and K

The most simple method of determining this relationship is to deduce the work which must be performed in a crack extending δa, as shown in Fig. 5[4]. If we consider a point A within δa, then the stress which was present before growth (when $r = \delta a - \xi$ and $\theta = 0$) will decrease to zero during growth. The displacement at A can be com-

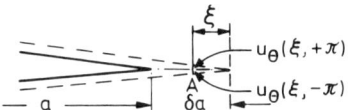

Fig. 5. Displacements during crack growth

puted from the new crack position (assuming that K_I does not change appreciably) since it is at $r = \xi$ and $\theta = \pm\pi$. The work done in unloading a small element of length $d\xi$ at A is therefore:

$$\frac{1}{2}\sigma_\theta(\delta a - \xi, 0)(u_\theta(\xi, +\pi) - u_\theta(\xi, -\pi)) B\, d\xi$$

so that the energy dissipated over the whole length δa is:

$$\delta U = -\frac{B}{2}\int_0^{\delta a} \sigma_\theta(\delta a - \xi, 0)(u_\theta(\xi, +\pi) - u_\theta(\xi, -\pi))\, d\xi$$

and:

$$G = \frac{1}{B}\frac{\delta U}{\delta a} = -\frac{1}{2\delta a}\int_0^{\delta a}\sigma_\theta(\delta a - \xi, 0)[u_\theta(\xi, +\pi) - u_\theta(\xi, -\pi)]\, d\xi$$

Substituting from Eq. (15) we have:

$$G_I = -\frac{1}{2}\int_0^{\delta a}\frac{K_I}{\sqrt{\delta a - \xi}}\left(-\frac{\sqrt{\xi}K_I 2(\kappa+1)}{8\pi\mu} - \frac{\sqrt{\xi}K_I 2(\kappa+1)}{8\pi\mu}\right)d\xi$$

which may be rearranged to give:

$$G_I = \frac{(\kappa+1)}{4\pi}\frac{K_I^2}{\mu}\frac{1}{\delta a}\int_0^{\delta a}\sqrt{\frac{\xi}{\delta a - \xi}}\, d\xi = \frac{(\kappa+1)}{8}\frac{K_I^2}{\mu}$$

For plane stress, this becomes:

$$G_I = \frac{K_I^2}{E} \tag{16}$$

so that G and K are interchangeable and a critical K_{IC} criteria is identical for linear elasticity to a critical G_c, which is correctly written as G_{IC}. In addition, K_I may be deduced from any G_I determination since:

$$K_I^2 = E\, G_I$$

so that both material properties are combined in K_I, a useful fact mentioned previously in Eq. (8).

2.5. Some Solutions for K_I

a) Cleavage Tests. As mentioned in Section 2.1., G may be determined from compliance measurements so that for the parallel cleavage specimen we had [Eq. (8)]:

$$G_c = P^2 \frac{12\, a^2}{B^2 D^3 E}$$

so that we may now write:

$$K_{IC}^2 = E\, G_{IC} = P^2 \frac{12\, a^2}{B^2 D^3}$$

and:

$$K_{IC} = 2\sqrt{3}\, \frac{P a}{B D^{3/2}} \tag{17}$$

Clearly, K_I is proportional to a and thus we have an inherently unstable situation with K_I increasing as the crack grows (see Section 4.5.). It would be useful to have a test which was stable in the sense that K_I did not change as the crack grew. This may be achieved by contouring the specimen such that $D^3 \propto a^2$, i.e. the depth of the

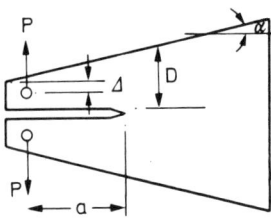

Fig. 6. Tapered cleavage specimen

specimen increases in a cubic form along its length. Precisely contoured cleavage specimens will achieve this but a simpler approximation[7] is a tapered beam, as shown in Fig. 6, in which:

$$D = D_0 + a \tan \alpha$$

In this case, D^3/a^2 remains approximately constant at $10.5 D_0$ for $\alpha \simeq 11°$ when $5 < a/D_0 < 20$, so that

$$K_{IC}^2 \simeq \frac{42 P^2}{B^2 D_0} \tag{18}$$

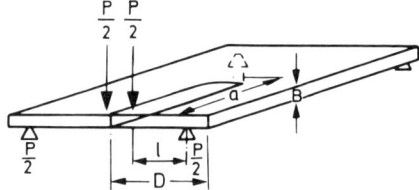

Fig. 7. The double torsion test

The invariance condition can be achieved by other geometries, providing that $C \propto a$ so that dC/da is a constant, and that of double torsion, as illustrated in Fig. 7, has proved useful for polymers[8]. In this case:

$$C = \frac{3(1+\nu) l^2 a}{E D B^3} + C_0, \quad (C_0 \text{ is } C \text{ at } a = 0)$$

so that:

$$K_{IC}^2 = \frac{3}{2} P^2 \frac{(1+\nu) l^2}{D B_c B^3} \tag{19}$$

at fracture. In this and other geometries, it is useful to put side grooves in the specimen to reduce the specimen thickness to B_c along the crack growth direction. In these cases, B in Eq. (9) should be replaced by B_c as in Eq. (19). An example of a calibration curve is given in Fig. 8[8] in which C has been determined experimentally for the double torsion test using PMMA specimens.

b) Stress Analysis Results. For some geometries, K_I may be conveniently deduced from the stress distribution for the whole body which may be derived by any of several methods[6]. The most useful is that of the infinite plate subjected to a uniform stress, σ, and containing a crack of length $2a$, as shown in Fig. 9. A stress function solution yields the result for σ_{yy} at $y = 0$[7]:

$$\sigma_{yy} = \frac{\sigma x}{\sqrt{x^2 - a^2}}$$

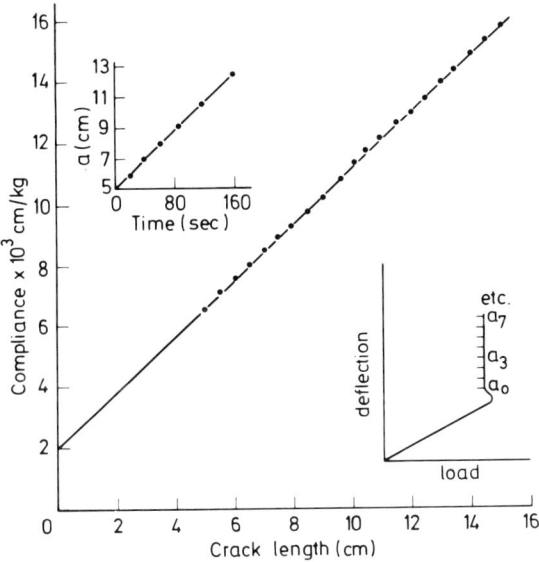

Fig. 8. Calibration curve for a double torsion specimen[8]
[Reproduced from Marshall, G. P., Coutts, L. H., and Williams, J. G.: J. Mats. Sci. 9, 1409 (1974).]

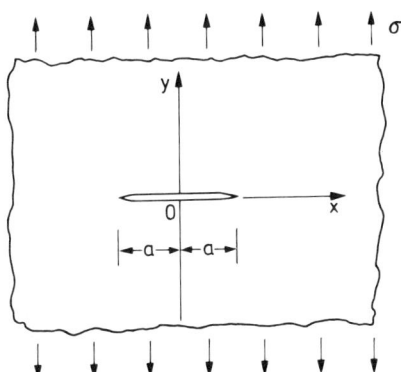

Fig. 9. Cracked infinite plate

Clearly, as $x \to \infty$, $\sigma_{yy} \to \sigma$ as expected and if $x = a + r$ the expression may be expanded, giving a singular term of the form:

$$\sigma_{yy} = \sigma \left(\frac{a}{2r} \right)^{1/2} + \ldots$$

Now, at $y = 0$, $\sigma_{yy} \equiv \sigma_{\theta\theta}$ at $\theta = 0$, so that, from Eq. (15), we have:

$$K_I = \sigma \sqrt{\pi a} \tag{20}$$

This is a most important result and is that deduced by Griffith, since it may be written as:

$$K_{IC}^2 = G_{IC}E = 2\gamma E = \sigma^2 \pi a$$

which is the Griffith Eq.[3]. The result is of particular value since it may be applied to all bodies in which there are small cracks which are loaded remotely from the crack. From purely dimensional arguments, the form of Eq. (20) may be deduced since σ and a are the only variables, and:

$$K_I = \text{constant} \cdot \sigma \sqrt{a}$$

For the infinite plate, the constant is $\sqrt{\pi}$, but for finite bodies the constant becomes a function of the various length parameters of the body. The equation is usually written as:

$$K_I^2 = Y^2 \sigma^2 a \tag{21}$$

and the function Y^2 has been tabulated for most practical geometries[9-11]. K_{IC} may be determined by plotting $Y^2 \sigma^2$ versus a^{-1} to give a straight line, the slope of which is K_{IC}^2. Figure 10 shows some data for PMMA plotted in this form, indicating the

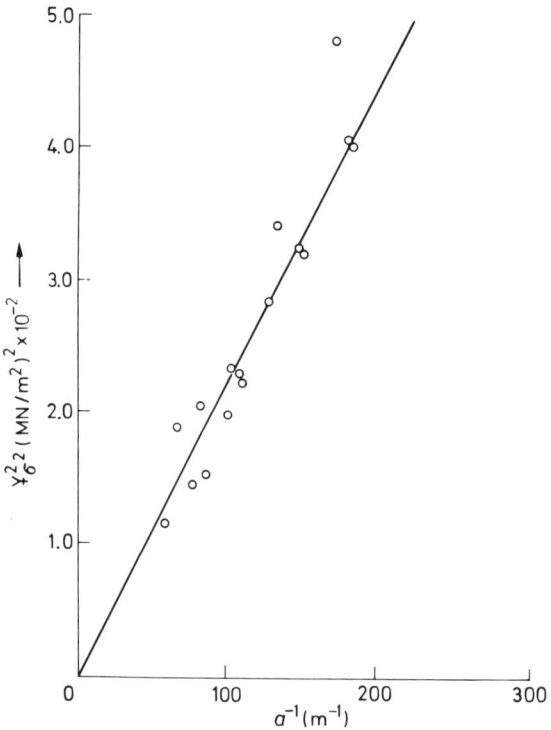

Fig. 10. Fracture data for PMMA at 20 °C
[Reproduced from Williams, J. G.: Polym. Engn. & Sci. *17* (3), 144 (1977)]

general level of scatter achieved in this sort of test. Y^2 is usually determined from numerical or analytical stress analyses but it is, of course, related to the compliance so that:

$$Y^2 = \frac{BD^2 E}{2a} \frac{dC}{da} \qquad (22)$$

The expression for the parallel cleavage test, for example, may be written in this form:

$$Y^2 = 12\left(\frac{a}{D}\right)$$

For the centre notched infinite plate, $Y = \sqrt{\pi}$, and for the edge notched wide plate, this becomes $1.12\sqrt{\pi}$, i.e. $\simeq 2$. For single-edge notched finite plates, we have[10]:

$$Y = 1.99 - 0.41\left(\frac{a}{D}\right) + 18.7\left(\frac{a}{D}\right)^2 - 38.48\left(\frac{a}{D}\right)^3 + 53.85\left(\frac{a}{D}\right)^4$$

for example, and similar polynomials are available for most geometries.

c) Energy Formulation. Equation (9) gives the equivalent form of the G_c relationship in terms of stored energy rather than load, and this can be written as:

$$U_1 = G_{IC} BD \phi \qquad (23)$$

where:

$$\phi = \frac{C}{dC/d(a/D)} \qquad (24)$$

ϕ may be expressed in terms of Y^2 from Eq. (22) so that:

$$\phi = \frac{\int Y^2 x \, da + BC_0/2}{Y^2 x} \qquad (25)$$

where $x = a/D$, and C_0 is the compliance for zero crack length.

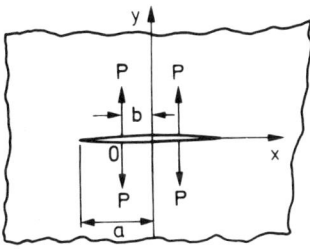

Fig. 11. Point loads on the crack faces

d) Point Loads. A result of special importance in subsequent sections is for the geometry shown in Fig. 11, in which point loads, P, are applied symmetrically to the crack faces so that K_I is given by[7]:

$$K_I = \frac{2P}{\sqrt{\pi}} \frac{\sqrt{a}}{\sqrt{a^2 - b^2}} \tag{26}$$

K_I for uniform pressure over the crack faces may be deduced by replacing P by $\sigma \, db$, so that this may be integrated to give:

$$K_I = \frac{2\sqrt{a}}{\sqrt{\pi}} \int_0^a \frac{\sigma \, db}{\sqrt{a^2 - b^2}} = \sigma \sqrt{\pi a} \tag{27}$$

This is the identical result to Eq. (20) for the stress at infinity and illustrates that, if a uniform pressure is applied to a cracked specimen, we have a stress intensity factor of $-\sigma \sqrt{\pi a}$ due to the external pressure and one of $+\sigma \sqrt{\pi a}$ due to that acting in the crack. Thus, the resultant value at the crack tip is zero.

3. Plastic Zones and Crazes

3.1. Small Scale Yielding

Since the stresses are singular at the crack tip, then clearly the yield criterion is exceeded in some zone in the crack tip region. If this zone is assumed to be small, then it will not greatly disturb the elastic stress field so that the extent of the plastic zone will be defined by the elastic stresses. If it is assumed that the Von Mises yield criterion is applicable (a reasonable first approximation for polymers), then the shape and size of the plastic zone may be derived from the stresses given in Eq. (15). Assuming a state of plane strain so that the transverse stress is given by $\nu(\sigma_{rr} + \sigma_{\theta\theta})$, then for a yield stress of σ_y, the plastic zone radius becomes:

$$r_p = \left(\frac{K_I^2}{2 \pi \sigma_y^2}\right) \cos^2 \frac{\theta}{2} \left(4(1 - \nu(1 - \nu)) - 3 \cos^2 \frac{\theta}{2}\right) \tag{28}$$

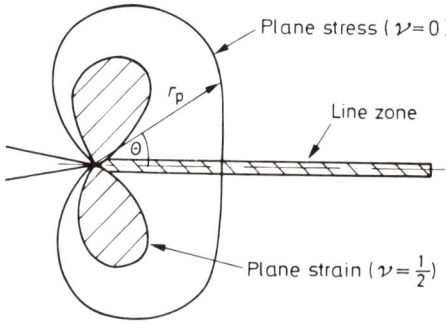

Fig. 12. Plastic zones

The shapes of the zones are shown in Fig. 12 for plane strain with $\nu = \frac{1}{2}$ and for $\nu = 0$ which is plane stress. The plane stress zone is the larger and is roughly semi-circular over the range $-\pi/2 < \theta < \pi/2$ with a radius of:

$$r_p = \frac{1}{2\pi} \frac{K_I^2}{\sigma_y^2} \tag{29}$$

The plane strain zone has two lobes with a maximum radius of $\frac{3}{4} r_p$. The smaller size is a consequence of the higher constraint in the plane strain case.

3.2. Line Zones and Crazes

In practice, the crack tip yielding in polymers is often not of a circular zone type as described above, but is a co-linear extension of the crack. The deformed material within the zone often forms a porous structure with ligaments restraining the zone faces, as illustrated in Fig. 13. This porous material, usually termed the craze, can be regarded as providing cohesive forces over the zone length. The zone can then be

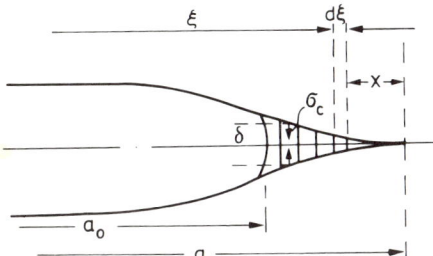

Fig. 13. Craze zone

modelled as a crack of length a with cohesive forces σ_c acting on the length $(a - a_0)$. The stress intensity factor may now be deduced for this case from Eq. (26) after the manner used in Eq. (27) so that:

$$K_I = \frac{2\sqrt{a}}{\sqrt{\pi}} \int_{a_0}^{a} \frac{(-\sigma_c) d\xi}{\sqrt{a^2 - \xi^2}}$$

Thus, there is a negative stress intensity factor, K_I, at the zone tip. If we now suppose that the zone will stabilize only when K_I is reduced to zero, then loads remote from the crack must be applied to produce an equal and opposite positive K_I value at the crack tip. Thus, the stable zone length is given by:

$$K_I = \frac{2\sqrt{a}}{\sqrt{\pi}} \int_{a_0}^{a} \frac{\sigma_c d\xi}{\sqrt{a^2 - \xi^2}} \tag{30}$$

where K_I is the applied stress intensity factor.

A result of special importance is that for small zones for which:

$r_p = a - a_0 \ll a$

so that Eq. (30) becomes:

$$K_I = \sqrt{\frac{2}{\pi}} \int_0^{r_p} \frac{\sigma_c \, dX}{\sqrt{X}}$$

where $X = a - \xi$.

For the special case in which σ_c is a constant, we have:

$$r_p = \frac{\pi}{8} \frac{K_I^2}{\sigma_c^2} \tag{31}$$

which may be compared with the plane stress zone size [Eq. (29)] showing that, for $\sigma_c = \sigma_y$, the line zone is about 2.5 times larger, as shown in Fig. 12. This solution is usually referred to as the Dugdale[12] model and has similarities with ideas put forward by Barenblatt[13].

The displacements within the zone[4] are given by:

$$\delta = \frac{(\kappa + 1) \sigma_c r_p}{\pi \mu} \left[\left(\frac{X}{r_p}\right)^{1/2} - \frac{1}{2}\left(1 - \frac{X}{r_p}\right) \ln\left(\frac{1 + (X/r_p)^{1/2}}{1 - (X/r_p)^{1/2}}\right) \right] \tag{32}$$

and at $X = 0$, the zone tip, $\delta = 0$. The maximum displacement, δ^*, at $X = r_p$ is given by:

$$\delta^* = \frac{(\kappa + 1) \sigma_c r_p}{\pi \mu}$$

which, for plane stress, becomes:

$$\delta^* = \frac{K_I^2}{E \sigma_c} = \frac{G_I}{\sigma_c}$$

i.e.

$$G_I = \delta^* \sigma_c \tag{33}$$

This result can, in fact, be deduced directly since the work done on any element of length dX which moves from the zone tip to the crack tip as the crack propagates is given by:

$$dU = \left(\int_0^{\delta^*} \sigma_c \, d\delta \right) dX$$

Since $dX = da$ during propagation:

$$G_I = \frac{dU}{da} = \int_0^{\delta^*} \sigma_c \, d\delta = \delta^* \sigma_c$$

for constant σ_c.

Side views of crazes show the cusp shape modelled here and precise measurements have been made of craze profile in transparent polymers using interference fringes[14]. Figure 14 shows some measurements compared with Eq. (32) made by Ward et al. which provide strong justification for a constant σ_c assumption.

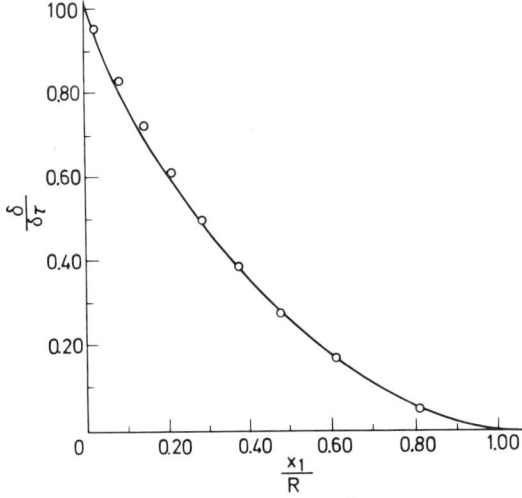

Fig. 14. Craze displacement profile[14]
[Reproduced from Morgan, G. P., and Ward, J. M.: Polymer 18, 87 (1977).]

A further use for this model is in providing a fracture criterion. Many processes in polymers are governed by strain rather than stress criteria as might be expected from consideration of molecular measurements, and for fracture both indirect[8] and direct[14] measurements confirm that a critical crack opening displacement (COD), δ_c^*, is a useful parameter for polymers. Thus, Eq. (33) may be written as a fracture condition:

$$G_{IC} = \delta_c^* \sigma_c$$

which is equivalent to a constant G_{IC} for constant σ_c and variations in G_{IC} are reflections of changes of σ_c, the cohesive or craze stress.

In some analyses, particularly of rubber modified materials[15], r_p may be quite large in practically interesting situations and some model of these cases in useful, especially when coupled with a COD criterion.

Solutions do exist for most geometries and they may be typified by that for an infinite plate giving [Eq. (20)]:

$$K_I = \sigma \sqrt{\pi a}$$

Combining this with Eq. (30) for a constant σ_c and integrating, we have:

$$\frac{r_p}{a_0} = \left(\sec \frac{\pi}{2} \frac{\sigma}{\sigma_c} - 1 \right) \qquad (34)$$

and the crack opening displacement for plane stress is:

$$\frac{\delta_c^*}{a_0} = \frac{8}{\pi} \frac{\sigma_c}{E} \ln \left(\sec \frac{\pi}{2} \frac{\sigma}{\sigma_c} \right) \qquad (35)$$

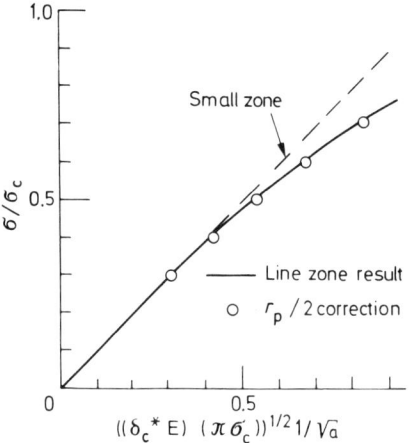

Fig. 15. Large plastic zone effects

Figure 15 shows this result plotted as σ/σ_c versus $[(\delta_c^* E)/(\pi \sigma_c)]^{1/2}(1/\sqrt{a_0})$ so that it may be compared with the conventional constant G_{IC} condition written in terms of δ_c^*, i.e.

$$\frac{\sigma}{\sigma_c} = \left(\frac{\delta_c^* E}{\pi \sigma_c} \right)^{1/2} \frac{1}{\sqrt{a_0}}$$

Substantial deviations from the small zone theory occur for $\sigma/\sigma_c > 0.5$ and some correction of this form must be employed under such circumstances.

A useful approximation to these plasticity results was originally suggested by Irwin[16] and assumes that, for small plastic zones, the crack length should be corrected by the addition of a length equal to $r_p/2$. The result is then precisely the same as the linear case, except that $a_0 + r_p/2$ replaces a_0 and we have:

$$G_{IC} = \delta_c^* \sigma_c = \frac{\pi \sigma^2 (a_0 + r_p/2)}{E}$$

and using Eq. (31) for r_p, we have:

$$\left(\frac{\sigma}{\sigma_c}\right)\left(1 - \frac{\pi^2}{16}\left(\frac{\sigma}{\sigma_c}\right)^2\right)^{-1/2} = \left(\frac{\delta_c^* E}{\pi \sigma_c}\right)^{1/2} \frac{1}{\sqrt{a_0}}$$

It should be noted that $r_p/2$ in Eq. (31) is approximately equal to r_p for the plane stress zone given by Eq. (29).

This result is also shown in Fig. 14, giving good agreement with Eq. (35) for which it is an approximation. The method is useful for making corrections in other geometries since it may be used in finite width correction factors.

3.3. Thickness Effects

In an edge notched plate specimen, the surface regions are in plane stress while the centre tends towards plane strain. The depth of the plane stress region is of the same order as the plane stress plastic zone radius given in Eq. (29), i.e.

$$r_p = \frac{1}{2\pi} \frac{K_I^2}{\sigma_c^2}$$

and the plate must be somewhat thicker than $2\,r_p$ to achieve plane strain in the centre. A useful working rule is that plane stress exists throughout the section for:

$$B < 4\,r_p = \frac{2}{\pi} \frac{K_I^2}{\sigma_c^2}$$

and, in general, this leads to through-section yielding and ductile fracture. For thicknesses greater than this, a plane strain region is established in the centre giving a fracture with plane stress edges and a plane strain centre. In metals, clear shear lips are formed over the full width of the plane stress region, but for polymers they are generally much smaller[17]. There is clear evidence of a thickness effect in K_{IC}, however, even when the fracture is flat indicating a difference between K_{IC} in the plane stress and plane strain stress states. This is most likely a reflection of a smaller δ_c^* in plane strain. A useful model of the mixed mode failure may be obtained by using two values of K_{IC}, i.e. K_{c1} in plane strain and K_{c2} in plane stress. For a mixed fracture, an average value is obtained given by[17]:

$$B K_{IC} = 2\,r_p K_{c2} + (B - 2\,r_p) K_{c1}$$

i.e.

$$K_{IC} = K_{c1} + \frac{2\,r_p}{B}(K_{c2} - K_{c1})$$

For $B < 4 r_p$, the value will become K_{c2} but, in general, this value is thickness dependent because of through-thickness yielding. For ductile fracture, $G_{IC} \propto B$ and we may write $K_{c2} \propto \sqrt{B}$. An overall view of the expected thickness dependence of K_{IC} on B is given in Fig. 16.

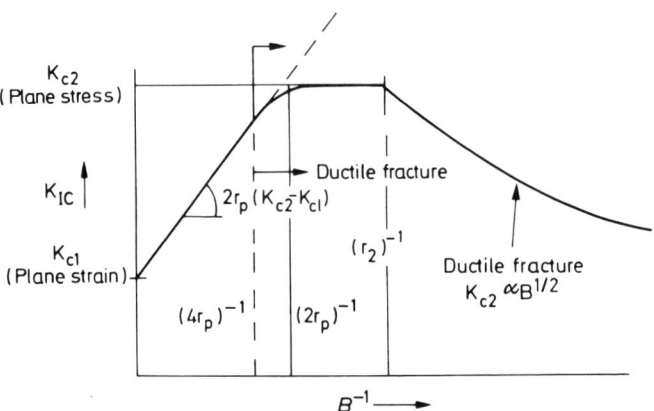

Fig. 16. Diagrammatic representation of K_{IC} as a function of thickness, B

The brittle polymers show very little evidence of thickness effects since r_p is small and, for practical thicknesses, the effective values are approximately K_{c1}. For tougher polymers in which r_p is large, the effect has been measured, for example, in polycarbonate[17], giving $K_{c2} \simeq 5$ MN/m$^{3/2}$ and $K_{c1} \simeq 2$ MN/m$^{3/2}$. For polyolefins, the effect is even more marked. Thin sheet fracture data has also been obtained[18].

4. Time Dependence

All the analysis so far described has assumed that the material is linearly elastic. Almost all polymers, however, show some evidence of time dependence and are viscoelastic. Since we are generally concerned with small strain behaviour for brittle cracks, it is reasonable to suppose that the materials are linearly viscoelastic so that, for example, when computing a strain from a stress, we cannot write:

$$e = \frac{1}{E} \sigma$$

but have:

$$e(t) = \int_0^t C(t - \tau) \frac{d\sigma(\tau)}{d\tau} d\tau \qquad (36)$$

where the convolution integral represents the time history effects embodied in the interaction between the stress input, $d\sigma(\tau)/d\tau$, and the creep compliance function, $C(t)$. Several attempts have been made to produce an exact representation of cracks in linearly viscoelastic materials[19-21] and notably in recent work by Schapery[22-24]. Restrictions on the type of behaviour described are necessary and useful results are only obtained by simplifications of the analysis, mostly by using particular material properties.

The analysis given here[8, 25] makes an assumption of this nature initially and derives similar results to the more complete formulations of the problem mentioned above. Most of the polymers considered in which brittle fracture occurs are only slightly viscoelastic and their time dependence may be expressed as:

$$C(t) = C_0 t^n$$

where n remains reasonably constant over a substantial range of rate and temperature. Thus, in creep, for example, we have:

$$e(t) = C_0 t^n \sigma$$

and n is the slope of a straight line on a log-log plot. n may be related approximately to the loss factor, $\tan \delta$, since, for a constant n:

$$\tan \delta = \tan \frac{\pi n}{2}$$

For most polymers, $n < 0.1$ and frequently $n < 0.05$. Under such circumstances, the convolution integral may be replaced by the time dependent equivalent elastic relationship so that:

$$e(t) = \frac{\sigma}{E(t)}$$

where:

$$E(t) = \frac{1}{C(t)}$$

Any errors incurred because of history effects will be of the order n, i.e. less than about 10%[7]. This assumption greatly simplifies the analysis since the linear elastic equations may be employed with the elastic parameters replaced by time dependent forms where the time scale is appropriate to the particular circumstances. In general, Poisson's ratio, ν, does not vary greatly and it is sufficient to use $E(t)$ and $\sigma_c(t)$ together with a constant COD criteria, δ_c^*. The most convenient method is to write:

$$K_{IC}^2 = E(t) G_{IC} = \delta_c^* E(t) \sigma_c(t) \tag{37}$$

and since K_I is a function of loading and geometry only, the time dependence of the fracture condition may be deduced.

4.1. Incubation Times

Suppose that a cracked body is loaded such that K_I is applied but that the crack does not grow since:

$$\delta^* = \frac{K_I^2}{E(0)\,\sigma_c(0)} < \delta_c^*$$

As time passes, both E and σ_c will decrease because of creep and eventually fracture will occur for $\delta^* = \delta_c^*$. In many polymers, σ_c is similar to the yield stress and the yield strain:

$$e_y = \frac{\sigma_c}{E}$$

remains constant with both time and temperature. Assuming a time dependence of the form:

$$E(t) = E_0\, t^{-n}$$

we have an expression for the time to initiate fracture of:

$$t_1 = \left(\frac{\delta_c^*\, e_y\, E_0^2}{K_{IC}^2}\right)^{1/2n} \tag{38}$$

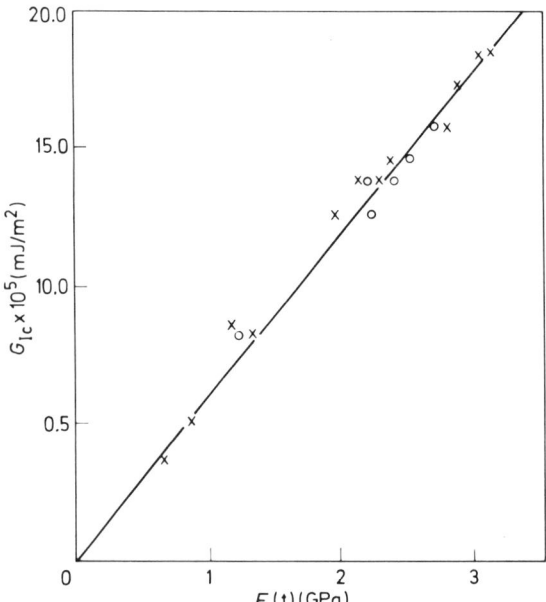

Fig. 17. G_{IC} as a function of time dependent modulus $E(t)$[26)]
[Reproduced from Gledhill, R. A., and Kinloch, A. J.: Polymer *17*, 727 (1976).]

Applications of Linear Fracture Mechanics

For a body containing small flaws of length a, the stress-time to fracture initiation relationship is given by:

$$\sigma = \left(\frac{\delta_c^* e_y E_0}{\pi a}\right)^{1/2} t_1^{-n} \tag{39}$$

An example of this effect is given in[26] where the time to failure of a polymeric adhesive is measured for various applied G_I values. The corresponding creep modulus values for the various times were then obtained. From Eq. (37):

$$G_{IC} = \delta_c^* \sigma_c(t) = (\delta_c^* e_y) E(t)$$

so that $G_{IC} \propto E(t)$ and this is shown to be the case in Fig. 17.

4.2. Slow Crack Growth

When crack growth occurs, the time scale in the craze zone changes from t, the elapsed time, to that determined by the speed of growth. The time taken for δ to increase from 0 to δ^* is:

$$t = \frac{r_p}{\dot{a}}$$

where $\dot{a}$ is the crack speed. Equation (37) now becomes:

$$K_{IC}^2 = (\delta_c^*)^{-n} e_y r_p E_0^2 \dot{a}^{2n}$$

i.e.

$$K_{IC} = \sqrt{\delta_c^* e_y} \left(\frac{8 e_y}{\pi \delta_c^*}\right)^n E_0 \dot{a}^n \tag{40}$$

Thus, there is a unique curve of K_{IC} versus $\dot{a}$ determined by the presence of a loss peak, giving a sufficiently high value of n to give stable crack growth (see Section 4.5. for a discussion of stability). Figure 18 shows a set of such curves for PMMA for a range of temperatures covering the β transition for which $n \simeq 0.07$. Such curves have been obtained on several polymers[8, 27–29] and indeed on a wide range of materials which show time dependence (e.g.[30, 31]). The most convenient experimental system is one in which K_I is independent of crack length so that a fixed crack speed may be established. The double torsion test described previously is ideal for this purpose.

The crack growth, or the time taken for a given system to reach some critical condition, may be deduced by integrating Eq. (40). For example, for a constant stress applied to a body containing a small flaw, crack growth will initiate at t_1 as given by Eq. (38). The speed at which the crack will start to move is:

$$\dot{a}_0 = \frac{r_p}{t_1}$$

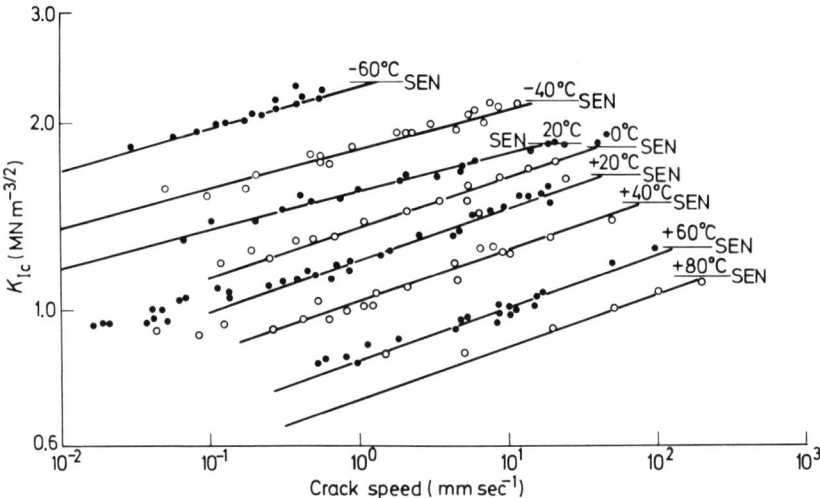

Fig. 18. K_{IC} as a function of crack speed[8] for PMMA
[Reproduced from Marshall, G. P., Coutts, L. H., and Williams, J. G.: J. Mats. Sci. 9, 1409 (1974).]

and it will continue to grow governed by the relationship:

$$\sigma \sqrt{\pi a} = \sqrt{\delta_c^* e_y} \left(\frac{8}{\pi} \frac{e_y}{\delta_c^*}\right)^n E_0 \dot{a}^n$$

This expression may be integrated to give the crack growth, Δa, at time t of the form:

$$\Delta a = r_p \left(\frac{t}{t_1} - 1\right) \tag{41}$$

for $\Delta a \ll a_0$. If we now suppose that unstable fracture occurs at $K_{IC} = K_{IC}^*$, a phenomenon which will be discussed in Section 4.5., then the time to reach final fracture may be computed, since:

$$K_{IC}^{*2} = \sigma^2 \pi (a_0 + \Delta a)$$

There is, of course, a critical crack speed corresponding to K_{IC}^*, $\dot{a}_c$, given by Eq. (40), and t_1 may be expressed as:

$$t_1 = \left(\frac{K_{IC}^*}{K_{I0}}\right)^{1/n} \frac{r_p}{\dot{a}_c}$$

where $K_{I0} = \pi \sigma^2 a_0$, the initial K_{IC} value. The parameter $r_p/\dot{a}_c$ is the limiting loading time below which no stable growth can ever occur.

Substitution in Eq. (41) yields the total time to failure under constant load:

$$t_2 = \left(\frac{r_p}{\dot{a}_c}\right)\left(\frac{K_{IC}^*}{K_{I0}}\right)^{1/n}\left\{\frac{a_0}{r_p}\left[\left(\frac{K_{IC}^*}{K_{I0}}\right)^2 - 1\right] + 1\right\} \qquad (42)$$

A similar argument may be applied to a constant rate of loading in which:

$$\sigma = R\,t$$

In this case, the initiation time becomes:

$$t_1 = \frac{r_p}{\dot{a}_c}\left[\frac{K_{IC}^*}{R\sqrt{\pi a_0}}\left(\frac{\dot{a}_c}{r_p}\right)\right]^{1/1+n}$$

and an initiation K_{IC} may also be deduced:

$$K_{I0} = \sqrt{\pi}\,R\,t_1\sqrt{a_0} = \left[R\sqrt{\pi a_0}\left(\frac{r_p}{\dot{a}_c}\right)\right]^{n/1+n} K_{IC}^{1/1+n} \qquad (43)$$

This result is for constant stress rate but, for small crack lengths, may be converted to strain rate by a simple modulus factor, i.e. $\dot{e} = R/E$. It should be noted that K_{I0} will be crack length dependent when determined at a fixed rate since $K_{I0} \propto a_0^{n/2(1+n)}$, but since $n \ll 1$ the dependence will not be strong. Figure 19 shows initiation K_{I0} values as a function of rate with K_{IC}^*, at instability, being rate independent.

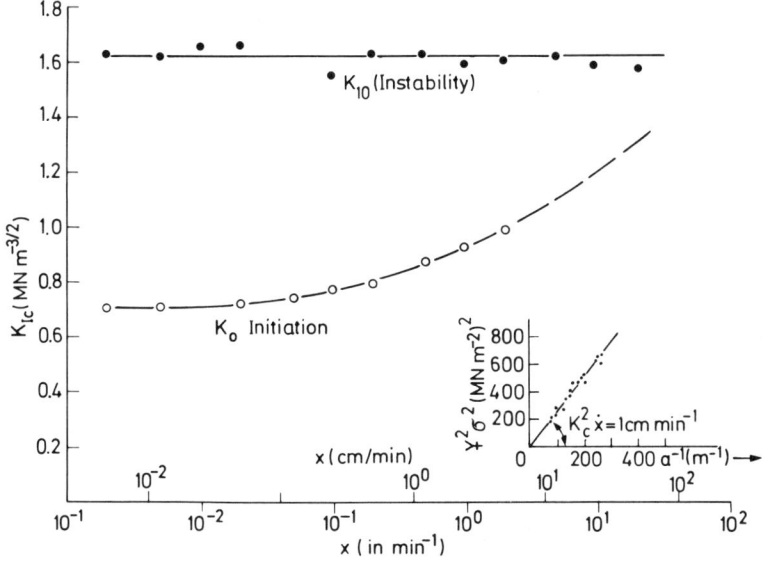

Fig. 19. Initiation K_{I0} as a function of rate[8] for PMMA
[Reproduced from Marshall, G. P., Coutts, L. H., Williams, J. G.: J. Mats. Sci. 9, 1409 (1974).]

The amount of slow growth before unstable fracture may also be deduced, and we have:

$$\Delta a = \frac{n}{1+n} r_p \left[\left(\frac{t}{t_1}\right)^{\frac{1+n}{n}} - 1 \right]$$

so that:

$$\Delta a = \frac{n}{1+n} r_p \left[\frac{K_{IC}^*}{R\sqrt{\pi a_0}} \left(\frac{\dot{a}_c}{r_p}\right) - 1 \right] \tag{44}$$

4.3. Craze Growth

During the incubation period of a crack, when δ^* increases, the length of the craze also increases. In unnotched specimens, there may be effects due to the nucleation of sites from which the crazes grow, but if an initial flaw is assumed, then from Eq. (31) we may write:

$$r_p(t) = \frac{\pi}{8} \frac{K_I^2}{\sigma_c^2(t)}$$

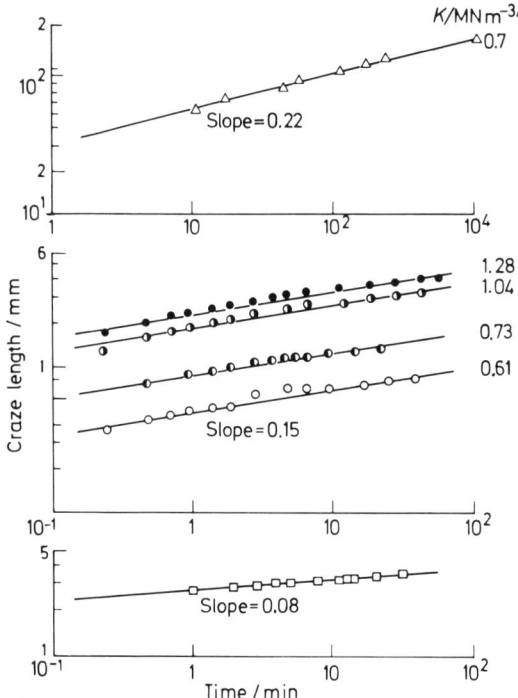

Fig. 20. Craze growth due to viscoelastic effects[32)]
[Reproduced from Williams, J. G., and Marshall, G. P.: Proc. R. Soc. A342, 55 (1975).]
(a) PMMA
(b) Rubber modified polystyrene
(c) Polycarbonate

Applications of Linear Fracture Mechanics

and using again the power law for convenience, we have:

$$r_p(t) = \frac{\pi}{8} \frac{K_I^2}{(e_y E_0)^2} t^{2n}$$

where:

$$t < t_1 = \left(\frac{\delta_c^* e_y E_0^2}{K_I^2}\right)^{\frac{1}{2}n}$$

The maximum length is when $t = t_1$ and is given by:

$$r_{p_{max}} = \frac{\pi}{8} \frac{\delta_c^*}{e_y}$$

Figure 20 shows data for growth of this type for three polymers[32].

4.4. Temperature Effects

The temperature dependence of modulus data can be quite accurately described using an Arrhenius equation with the concept of time-temperature equivalence. An equivalent time, t', may thus be deduced for a process at an absolute temperature, T, which occurs at t for an original temperature, T_0, such that:

$$t' = t \, e^{-(H/R)(1/T - 1/T_0)} \tag{45}$$

where R is the gas constant and H is the activation energy for a given molecular process. Values of H for the various viscoelastic processes in polymers (α, β, γ, etc.) are well established[33], making Eq. (45) a convenient representation. If the modulus $E(t)$ is described in this way, then Eq. (40) becomes:

$$K_I = \sqrt{\delta_c^* e_y} \left(\frac{8}{\pi} \frac{e_y}{\delta_c^*}\right)^n E_0 \, e^{n(H/R)(1/T - 1/T_0)} \, \dot{a}^n \tag{46}$$

where n corresponds to the particular viscoelastic process. It should be noted that n will only be constant over the central part of the temperature range of the process and will decrease both above and below.

4.5. Crack Stability

A crack will be termed stable here, providing that:

$$\frac{d\dot{a}}{\dot{a}} \leq 0 \tag{47}$$

i.e. it moves at constant speed or decelerates. For elastic materials with $n = 0$, the crack growth is always unstable, but for the general case from Eq. (40), we have:

$$\frac{d\dot{a}}{\dot{a}} = \frac{1}{n} \frac{dK_I}{K_I} \qquad (48)$$

For most specimen geometries, $dK_I/K_I \geqslant 0$ so that cracks accelerate and are unstable. For the specially designed constant K_I specimens, $dK_I = 0$ so that constant speed growth is obtained. However, there are always errors and inaccuracies in every practical test and dK_I/K_I represents the inherent accuracy of these tests. For example, a 1% load variation with $n = 0.05$ will give a change of crack speed of 20% from Eq. (48). This sort of acceleration is easily controlled but, if n decreases to 0.01, then the speed would double and the specimen becomes impossible to control. It seems likely that, with practically achievable accuracies of testing, values of $n > 0.01$ would be necessary to achieve stable crack growth. This stable growth can only be achieved within the range of a viscoelastic transition. This effect can be observed when cracks are grown at quite low speeds but become unstable at both high and low temperatures.

A similar effect would be expected therefore at high and low speeds for a given temperature and at low speeds this is the case. Such transitions follow the activation energy of the viscoelastic process involved[8, 27] but, at high speeds, crack instability occurs which does not, and indeed occurs when n has a substantial value. The most likely explanation of this, the most important instability which determines K_{IC}^*, is that an isothermal-adiabatic transition occurs in the craze region such that the heat generated in the craze is not conducted away. This results in thermal softening in the craze with a consequent decrease in K_{IC} and a resulting instability.

The maximum temperature rise in a craze of thickness b is given by:

$$\Delta \widetilde{T} = \frac{G_{IC}}{\rho c b} \qquad (49)$$

where ρ is the density, and c is the specific heat.

If we now consider the craze as a uniform strip with heat generated at a fixed rate, then the temperature rise in the strip at time t is given by[34]:

$$\Delta T = \Delta \widetilde{T} \left(1 - 4 i^2 \operatorname{erf} c \sqrt{\frac{b^2 \rho c}{16 k t}}\right)$$

where k is thermal conductivity, and $i^2 \operatorname{erf} c\, x$ is the double integral of $(1 - \operatorname{erf} x)$ where $\operatorname{erf} x$ is the error function. b is known to be small for crazes since it is of the order δ^* so that the error function may be expanded to give[8]:

$$\Delta T = \frac{\delta_c^* e_y E(t)}{\sqrt{\pi \rho c k}} \frac{1}{\sqrt{t}} = \frac{e_y}{\sqrt{\rho c k}} K_{IC} \sqrt{\dot{a}} \qquad (50)$$

so that:

$$\frac{dT}{d\dot{a}} = (T - T_0)\left(\frac{1}{K_{IC}} \frac{dK_{IC}}{d\dot{a}} + \frac{1}{2\dot{a}}\right)$$

and the rate of change of temperature with crack speed in the craze is known.

By substituting Eq. (50) in Eq. (46), the coupling effect of rate and temperature rise on the $K_{IC} - \dot{a}$ curve can be described but since we are interested here in the stability condition, we may differentiate Eq. (46) to give:

$$\frac{1}{K_{IC}} \frac{dK_{IC}}{d\dot{a}} = n\left(\frac{1}{\dot{a}} - \frac{H}{RT^2} \frac{dT}{d\dot{a}}\right)$$

and substituting for $dT/d\dot{a}$, we have:

$$\frac{d\dot{a}}{\dot{a}} = \frac{1}{n} \frac{dK_{IC}}{K_{IC}} \left(\frac{1 + n(H/R)\,\Delta T/(T_0 + \Delta T)^2}{1 - (H/2R)\,\Delta T/(T_0 + \Delta T)^2}\right) \tag{51}$$

An alternative instability condition of $n \to 0$ is now apparent since $d\dot{a}/\dot{a} \to \infty$ when:

$$\frac{H}{2R} \frac{\Delta T}{(T_0 + \Delta T)^2} = 1$$

i.e.

$$\frac{2RT_0}{H} = \frac{\Delta T/T_0}{[1 + (\Delta T/T_0)]^2}$$

For most polymers $H/2R \gg T_0$, so that this may be approximated to:

$$\Delta T = \frac{2RT_0^2}{H} \tag{52}$$

and the instability crack speed may be deduced from Eq. (50) as:

$$\dot{a}_c = \frac{\rho c k}{(e_y K_{IC}^*)^2} \left(\frac{2R}{H}\right)^2 T_0^4 \tag{53}$$

Since K_{IC}^* is also a function of T_0, we may compare with experimental values by plotting $\dot{a}_c K_{IC}^{*2}$ versus T_0^4, as shown in Fig. 21 for PMMA (from[8]). The expected linear dependence is apparent and, indeed, the various thermal parameters for PMMA provide a reasonable estimate of the slope of the line.

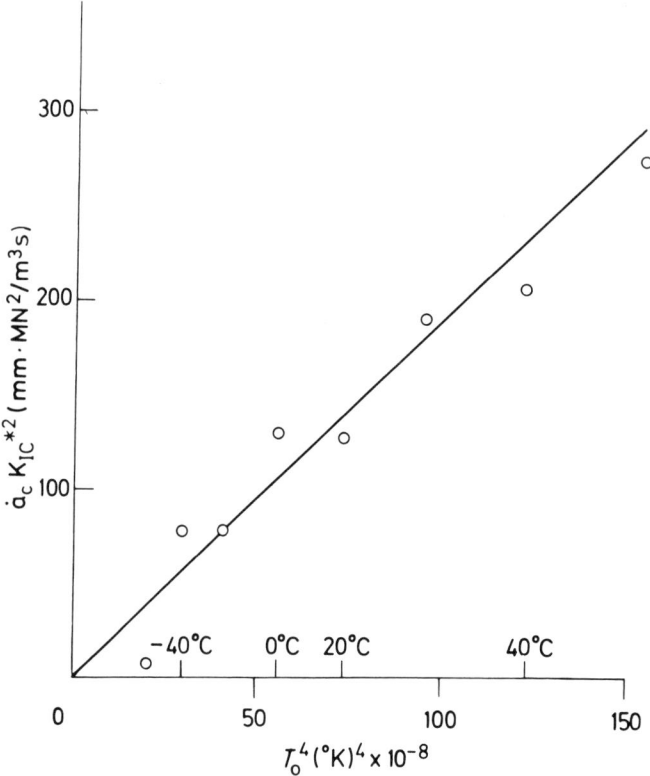

Fig. 21. Instability crack speeds for PMMA as a function of temperature

5. Environmental Effects

The rôle of environments in the stress enhanced cracking and crazing of polymers has stimulated much debate in recent years, but it seems likely that a mechanism involving the plasticisation of the crack tip craze is the best description. Many fluids which crack or craze polymers are chemically inert with them and diffuse into the bulk only slowly. The craze, however, because of its porous nature, has a very high area to volume ratio so that penetration of the fluid only a small distance leads to a complete plasticisation of the ligaments and a subsequent drop in their load carrying capacity, σ_c. Since the ligaments are very small ($\simeq 200$ Å), this occurs very rapidly so that the material behaves almost instantaneously as one with a lower craze stress. This mechanism may be modelled very simply using the line model by simply replacing σ_c by $\alpha \sigma_c$. For convenience, we will assume that the viscoelastic effects in the craze remain unaltered, although they are often greatly reduced by plasticisation. Those of the bulk material embodied in E will not, of course, be affected.

5.1. Crack Growth

If we now assume that cracking occurs at the same value of δ_c^*, then the critical K_{IC} value is given by:

$$K_{IC} = \sqrt{\delta_c^* e_y} E \sqrt{\alpha}$$

i.e. the critical value has been reduced by a factor of $\sqrt{\alpha}$. In time dependent behaviour, Eq. (40) now becomes:

$$K_{IC} = \sqrt{\delta_c^* e_y} E_0 \sqrt{\alpha} \left(\frac{8 e_y \alpha}{\pi \delta_c^*} \right)^n \dot{a}^n \tag{54}$$

In Fig. 22, this expression is shown together with that for air (assumed inert) and it can be seen that the environmental curve is simply that for air displaced to lower K_I values and probably with a reduced slope.

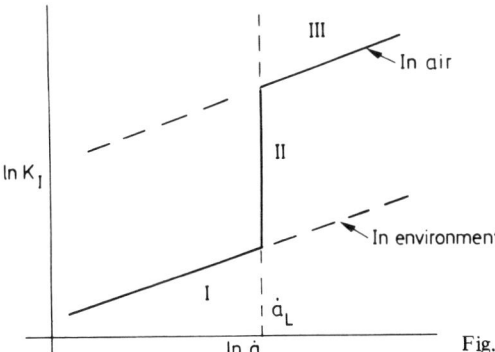

Fig. 22. Environmental crack growth

This behaviour assumes that the environment is instantaneously available within the craze and clearly at high speeds this will not be so. The limiting factor will be the flow of the fluid within the craze and we would expect a transition from the lower to the upper line as crack speed increases. An estimate of the effect may be made by assuming that the flow is governed by D'Arcy's Law for flow in porous media such that the flow velocity, V, is given by:

$$V = \frac{A}{12 \mu} \left(\frac{dp}{dx} \right) \tag{55}$$

where A = the pore area
dp/dx = some form of pressure gradient
and μ = the fluid viscosity

A may be modelled[32] as being proportional to δ^* so that in crack growth it is constant since $\delta^* = \delta_c^*$. Assuming an approximately constant pressure gradient, arising from atmospheric pressure and surface tension, we would expect a limiting velocity resulting in a limiting crack speed, say $\dot{a}_L$. The K_{IC} versus $\dot{a}$ curve, therefore, has three distinct portions, as shown in Fig. 22.

Within the transition region II, it is possible to achieve equilibrium within the craze zone since there can exist a two stage zone in which part is plasticised to $\alpha \sigma_c$

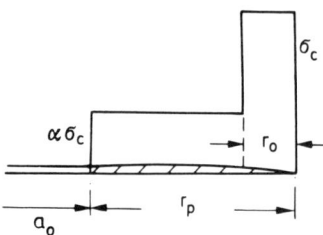

Fig. 23. Two stage zone

and the tip is at σ_c, as shown in Fig. 23. For an applied value of K_I, we may use Eq. (30) to give:

$$K_I = \frac{2\sqrt{a+r_p}}{\sqrt{\pi}} \int_{a_0}^{a_0+r_p} \frac{\alpha \sigma_c \, d\xi}{\sqrt{a^2 - \xi^2}} + \frac{2\sqrt{a+r_p}}{\sqrt{\pi}} \int_{a_0+r_p-r_0}^{a_0+r_p} (1-\alpha) \frac{\sigma_c \, d\xi}{\sqrt{a^2 - \xi^2}}$$

Assuming that r_p and $r_0 \ll a_0$, then we may write:

$$K_I = \sqrt{\frac{2}{\pi}} \int_0^{r_p} \frac{\alpha \sigma_c \, dX}{\sqrt{X}} + \sqrt{\frac{2}{\pi}} \int_0^{r_0} (1-\alpha) \sigma_c \frac{dX}{\sqrt{X}}$$

i.e.

$$K_I = 2\sqrt{\frac{2}{\pi}} \sigma_c [\alpha \sqrt{r_p} + (1-\alpha)\sqrt{r_0}] \qquad (56)$$

For any value of K_I, therefore, a range of r_p values is possible:

$$\frac{\pi}{8} \frac{K_I^2}{\sigma_c^2} \leq r_p \leq \frac{\pi}{8} \frac{K_I^2}{\alpha^2 \sigma_c^2}$$

For the same conditions, we may deduce a displacement, δ^*, for $r_0/r_p \ll 1$:

$$\delta^* = \frac{8}{\pi} \frac{\sigma_c}{E} [\alpha r_p + 2(1-\alpha)\sqrt{r_p r_0}] \qquad (57)$$

Thus, equilibrium is achieved with $\delta^* = \delta_c^*$ and $\dot{a} = \dot{a}_L$ for any K_I value. In fact, a transition to stage III will occur at higher K_I values since a higher speed equilibrium may be achieved without the environment being present. These concepts seem to be

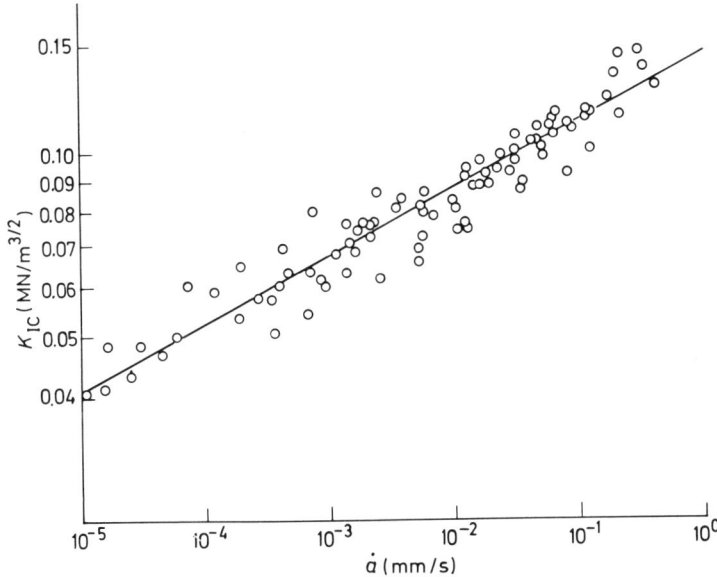

Fig. 24. K_{IC}-crack speed data for polyethylene (MFI 20) at 20 °C
[Reproduced from Williams, J. G.: Polym. Engn. & Sci. *12* (1), 85 (1976).]

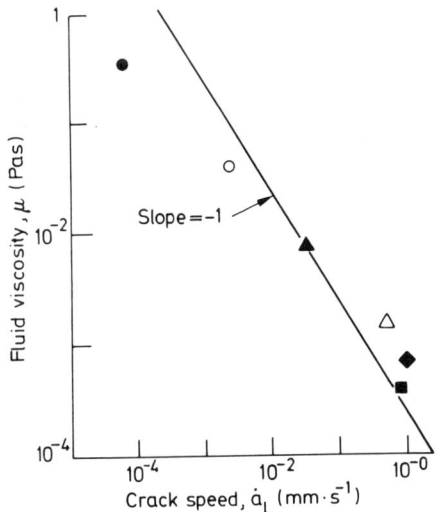

Fig. 25. Viscosity effects in environmental crack growth[32)]
[Reproduced from Williams, J. G., and Marshall, G. P.: Proc. R. Soc. *A342*, 55 (1975).]
HDPE in detergent at ○ 20 °C,
● 60 °C   Rubber modified polystyrene
in ▲ paraffin oil, △ white spirit,
■ n-heptane, ◆ LDPE in methanol –
from Fig. 24

basically correct but there is little supporting evidence in the literature. Figure 24 shows a stage I environmental curve for polyethylene in alcohol ($n \simeq 0.11$), indicating an upper limit speed, $\dot{a}_L$, of about 1 mm/s. Two polymers with liquids of various viscosities were used to determine $\dot{a}_L$ and this is shown as a function of μ in Fig. 25[32)]. The K_{IC} values are roughly the same for the various materials so that the inverse proportionality between μ and $\dot{a}_L$ is in accordance with Eq. (55).

5.2. Craze Growth

A similar line of argument applies to craze growth in environments which can occur for $\delta^* < \delta_c^*$. At very slow speeds (at long times), the environment is always available and we have craze growth of the same form as in air, except that the craze stress is now $\alpha \sigma_c$ so crazes are now generally longer by a factor of α^{-2}. The time dependent processes controlling the growth are those of $\alpha \sigma_c$ and are likely to be affected by the plasticisation process. Since this is a swelling process which essentially makes the material behave as if it is at its glass transition (*i.e.* become rubbery)[35], then most time dependent processes would be expected to be greatly reduced so that there will be very little long term growth.

At shorter times, the growth process again is controlled by the availability of the environment but now there is no single limiting speed since $\delta^* \neq \delta_c^*$. The pore area may be approximated by assuming that the craze strain is given by δ^*/d_0, where d_0 is a measure of craze thickness, and if it is assumed that pores form at a spacing of l_0, then we have[32, 36]:

$$A = \frac{l_0^2 \, \delta^*}{2 d_0}$$

Subsequent swelling may change l_0 and d_0 but it is assumed here that the pore size is governed by the original value of δ^* before environmental effects occur. The pore area now becomes:

$$A = \frac{l_0^2}{2 d_0} \frac{K_I^2}{E \sigma_c}$$

and the craze velocity is given by Eq. (55) since $V = dr_p/dt$:

$$\frac{dr_p}{dt} = C K_I^2 \left(\frac{dp}{dx} \right) \tag{58}$$

where:

$$C = \left(\frac{l_0^2}{24 \, d_0 \, E \, \sigma_c \, \mu} \right),$$

a constant for any system.

The most simple system in which to observe this flow controlled growth is in a craze grown from a crack such that the flow is simply directed along the craze length. If it is assumed that the pressure in the newly formed voids at the craze tip is zero and that at the crack tip the pressure is $\bar{p}$ (generally atmospheric plus some capillary force contribution), then:

$$\frac{dp}{dx} = \frac{\bar{p}}{r_p}$$

Substituting in Eq. (58) and integrating gives a growth law of the form:

$$r_p = \sqrt{2 C \bar{p}} \, K_I \, t^{1/2} \tag{59}$$

(For $r_p \gg r_0$, the original value.) Figure 26 shows some data for crazes grown in PMMA immersed in methanol at various K_I values showing the $t^{1/2}$ relationship[32].

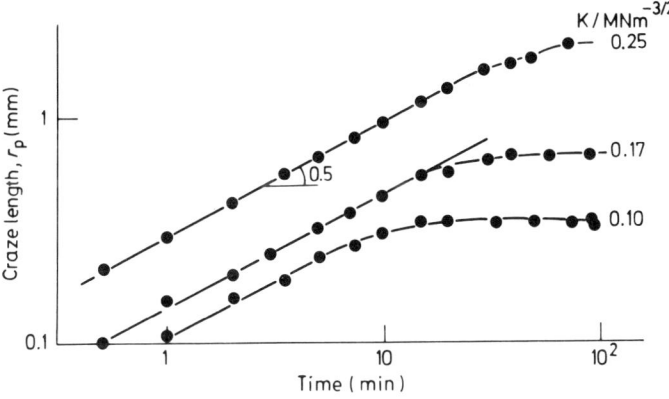

Fig. 26. Craze growth in an environment[32] – PMMA in methanol
[Reproduced from Williams, J. G., and Marshall, G. P.: Proc. R. Soc. *A342*, 55 (1975).]

The crazes are seen to arrest and remain stationary after some time and show little sign of subsequent relaxation controlled growth. Arrest should occur when:

$$\hat{r}_p \to \frac{\pi}{8} \frac{K_I^2}{\alpha^2 \sigma_c^2}$$

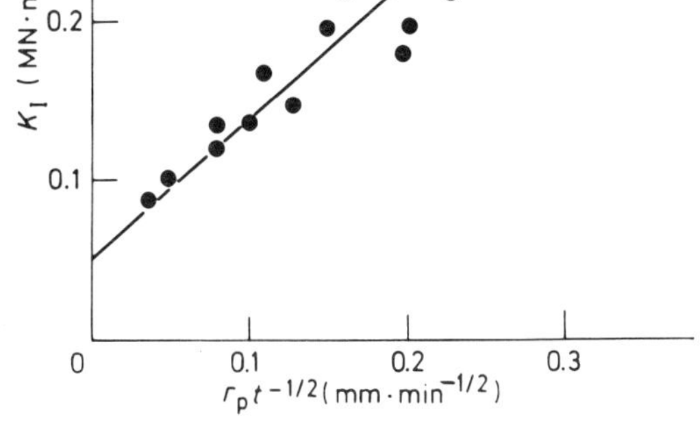

Fig. 27. The dependence of craze growth on stress intensity factor, K_I[32] – PMMA in methanol
[Reproduced from Williams, J. G., and Marshall, G. P.: Proc. R. Soc. *A342*, 55 (1975).]

but in fact it occurs at lengths much less than this. The most likely explanation appears to be the accumulation of debris which blocks the craze pores, which occurs at a fixed time, say $\bar{t}$, for all K_I values. The approximate time for the PMMA-methanol system is 20 minutes. The dependence on K_I is demonstrated in Fig. 27 in which the K_I values are plotted versus $r_p/t^{1/2}$ giving a straight line as predicted. An initiation value of K_I, K_m, is indicated which probably represents a minimum strain to form pores.

The growth of flow controlled crazes for other geometries may be predicted from the two-dimensional version of the flow equation in which:

$$u = C K_I^2 \left(-\frac{\partial p}{\partial x} \right) \quad \text{and} \quad v = C K_I^2 \left(-\frac{\partial p}{\partial y} \right)$$

where u und v are the velocities of flow in the x and y directions, respectively. From continuity, we have:

$$\frac{\partial u}{\partial x} + \frac{\partial v}{\partial y} = 0$$

so that the pressure distribution must satisfy the Laplace equation:

$$\nabla^2 p = \frac{\partial^2 p}{\partial x^2} + \frac{\partial^2 p}{\partial y^2} = 0 \tag{60}$$

Some special cases are worthy of consideration:

(i) One-dimension: This is the case already discussed in which Eq. (60) becomes:

$$\frac{d^2 p}{dx^2} = 0 \quad \text{and} \quad p = A x + B$$

The boundary conditions are $p = \bar{p}$ at $x = 0$ and $p = 0$ at $x = x_0$, so that:

$$p = \bar{p} \left(1 - \frac{x}{x_0} \right)$$

The craze speed is given by u at $p = 0$, so that:

$$\frac{dp}{dx} = -\frac{\bar{p}}{x_0}$$

and we have:

$$\frac{dx_0}{dt} = C K_I \frac{\bar{p}}{x_0}$$

as in Eq. (58).

(ii) Axial Symmetry: For crazes grown inwards from circumferentially notched bars or outwards from internal penny-shaped cracks, the pressure distribution must be axi-symmetric so that:

$$\frac{d^2p}{dr^2} + \frac{1}{r}\frac{dp}{dr} = 0$$

i.e.

$$p = A \ln r + B$$

For a craze growing inwards from a crack at tip radius a to a distance x_0, we have:

$$p = \bar{p} \text{ at } r = a \quad \text{and} \quad p = 0 \text{ at } r = a - x_0$$

and the pressure gradient at $p = 0$ is:

$$\left(\frac{dp}{dr}\right)_0 = -\frac{\bar{p}}{(a - x_0)\ln(1 - x_0/a)}$$

and the growth is given by:

$$\frac{dx_0}{dt} = \frac{C K_I^2 \bar{p}}{a(1 - x_0/a)\ln(1 - x_0/a)}$$

This may be integrated to give:

$$\frac{1}{2}\left(1 - \frac{x_0}{a}\right)^2 \left(\ln\left(1 - \frac{x_0}{a}\right)^2 - 1\right) + \frac{1}{2} = \frac{2 C \bar{p} K_I^2}{a^2} t$$

i.e.

$$\phi\, x_0 = \sqrt{2 C \bar{p}}\, K_I\, t^{1/2} \tag{61}$$

where:

$$\phi = \frac{\{(1 - x_0/a)^2 (\ln(1 - x_0/a)^2 - 1) + 1\}^{1/2}}{\sqrt{2}\, x_0/a}$$

is the necessary correction factor for the circular geometry. Fig. 28 shows some data, again for PMMA in methanol, taken from[37] in which craze growth data is shown plotted as x_0 versus $\sqrt{t}$ both with and without ϕ, and it is clear that the correction factor is effective. Inward growing crazes tend to accelerate, as would be expected, while outward growing crazes would decelerate. The equivalent correction factor for this case is obtained, replacing $(1 - x_0/a)$ with $(1 + x_0/a)$ in Eq. (61).

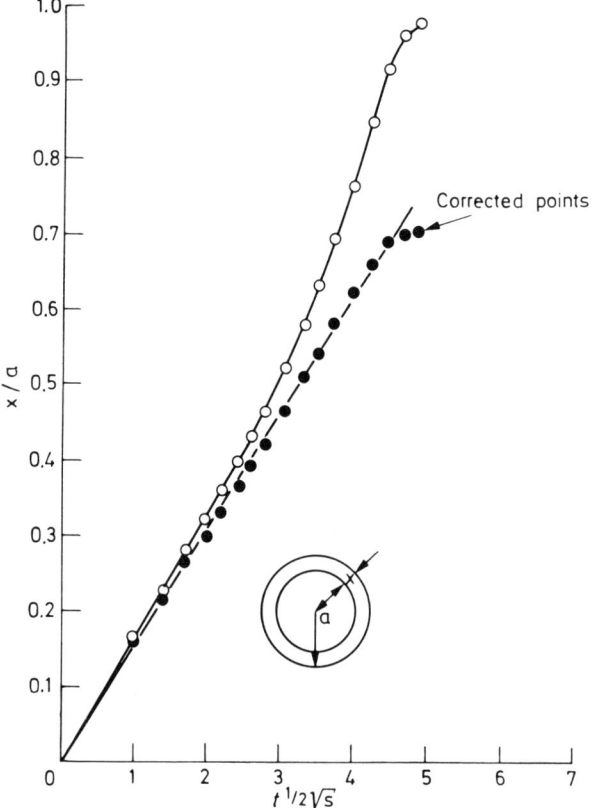

Fig. 28. Axi-symmetric craze growth (data from[37]) – PMMA in methanol

(iii) Growth on a Free Surface: When crazes grow along a free surface, the flow of the environment can be augmented by flow through the surface. At the junction of the craze front and the surface, there is theoretically a dual boundary condition of $p = 0$ and $p = \bar{p}$ which would result in infinite rates near the surface which do not occur. There appears to be two limiting conditions at this interface, namely:
a) the K_I value must exceed some critical value; and
b) equal flow rates occur both along and transverse to the free surface.

If we consider a craze grown from some small initial flaw which has propagated along the surface, a profile of the form shown in Fig. 29 would be expected. The pressure profile within the craze must satisfy Eq. (60) and, in general, terms of the form:

$e^{\lambda y}(\sin \lambda x, \text{ or } \cos \lambda x)$

are useful. For the coordinate system shown in Fig. 29, the distribution can take the form:

$$p = \bar{p}\left(1 - \frac{x}{\bar{x}_0} - e^{-\lambda y} \cos \lambda x\right) \tag{62}$$

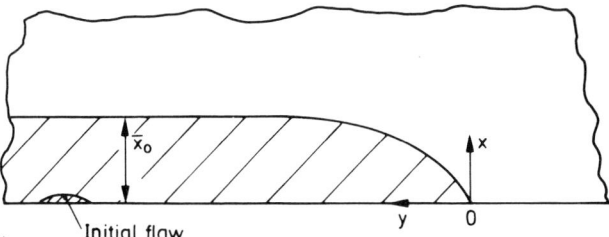

Fig. 29. Craze growth along a free surface

where $\bar{x}_0$ is the craze depth remote from the propagating front, and λ is a constant. Note that for large y the one-dimensional result is recovered. At the craze-boundary interface of $x = y = 0$, if we assume $u = v$, then $\partial p/\partial x = -\partial p/\partial y$, and:

$$\lambda = \frac{1}{\bar{x}_0}$$

The craze boundary is given by $p = 0$ so that:

$$e^{y_0/\bar{x}_0} = \frac{\cos x_0/\bar{x}_0}{1 - x_0/\bar{x}_0}$$

and the velocities are:

$$\frac{dx_0}{dt} = C K_I^2 \frac{\bar{p}}{\bar{x}_0}\left(1 - \tan\frac{x_0}{\bar{x}_0}\left(1 - \frac{x_0}{\bar{x}_0}\right)\right)$$

and:

$$\frac{dy_0}{dt} = C K_I \frac{\bar{p}}{\bar{x}_0}\left(1 - \frac{x_0}{\bar{x}_0}\right)$$

The penetration of the craze, $\bar{x}_0$, is constant for $t > \bar{t}$, so that the velocity along the surface at $x_0 = 0$ is a constant for a given K_I, and given by:

$$\left(\frac{dy_0}{dt}\right) = \sqrt{2 C \bar{p}}\,\frac{K_I}{2\sqrt{\bar{t}}} = \frac{\bar{x}_0}{2\bar{t}} \tag{63}$$

i.e. the craze penetration velocity at arrest.

(iv) Growth from an Edge Notch: A special case of the free surface case is when there are two free surfaces close together, as in the case of a craze growing from an edge notch, as shown in Fig. 32. In this case, there is a symmetrical pressure distribution given by:

$$p = \bar{p}(1 - e^{-\lambda y}\cos \lambda x)$$

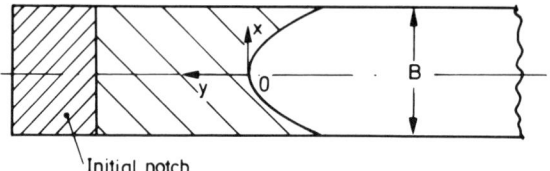

Fig. 30. Craze growth from an edge notch

At $x = B/2$, $\partial p/\partial x = \partial p/\partial y$ as before, so that:

$$\sin \frac{\lambda B}{2} = \cos \frac{\lambda B}{2}$$

which is achieved by $\lambda = \pi/2B$. The craze front profile is again given by $p = 0$:

$$e^{\pi y_0/2B} = \cos \frac{\pi x_0}{2B}$$

and the velocity along the sheet is:

$$\frac{dy_0}{dt} = C K_I^2 \bar{p} \frac{\pi}{2B} \tag{64}$$

i.e. a constant proportional to K_I^2 and inversely proportional to thickness.

Such constant speed growth is observed at higher K_I values in the PMMA-methanol system[36] and in some others[32, 38]. Figure 31 shows the craze velocity as a func-

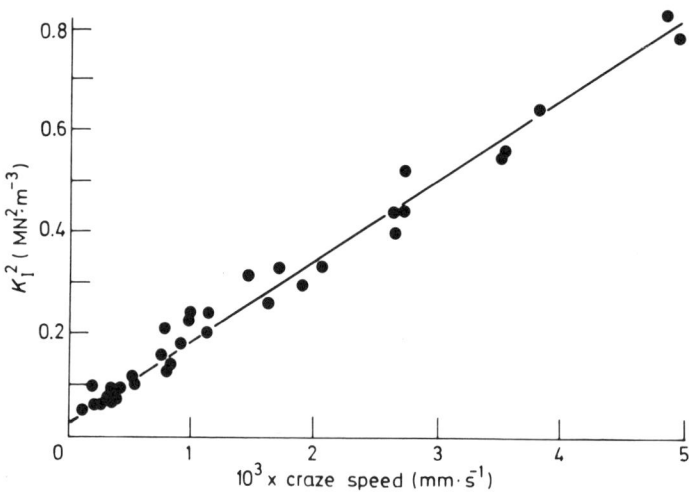

Fig. 31. Dependence of constant speed craze growth on K_I[32] – PMMA in methanol
[Reproduced from Williams, J. G., Marshall, G. P.: Proc. R. Soc. *A342*, 55 (1975).]

tion of K_I^2 giving the expected linearity. The inverse thickness dependence is also observed[36] and, by using the $t^{1/2}$ data on the same system, the proportion of B can be determined. The experiments indicate a value of 0.76 compared with the theoretical value of $2/\pi = 0.64$.

6. Fatigue

Crack growth in polymers is induced by cyclic loading as in most other materials. When crack growth from an existing crack is observed, it is found that the well known Paris law works well:

$$\frac{da}{dN} = (\Delta K_I)^m$$

where da/dN = crack growth per cycle, ΔK_I = range of stress intensity factor, and m is a constant of approximately four. This law has been extensively studied (e.g.[39-41]), especially with regard to mean stress and frequency effects, and there is a good deal of data available on a wide range of polymers. No details will be given here since an excellent and comprehensive review[42] is available which gives most of the available information.

A point worthy of mention here, however, is that a close analogy may be drawn with fatigue and environmental effects since the effect of cycling is to reduce σ_c to, say, $\alpha \sigma_c$. If a very simple model is used, then a single cycle can affect this reduction, so that if K_I is applied, the original zone length is:

$$r_{p_0} = \frac{\pi}{8} \frac{K_I^2}{\sigma_c^2}$$

After one cycle this zone can now only sustain $\alpha \sigma_c$ so that a two-stage zone is established, as in Fig. 23. In this zone, the new length, r_{p_1}, is given by Eq. (56) such that:

$$K_I = 2\sqrt{\frac{2}{\pi}} \sigma_c \left[\alpha \sqrt{r_{p_1}} + (1 - \alpha) \sqrt{r_0} \right]$$

and we know that:

$$r_{p_1} - r_0 = r_{p_0}$$

If the process is repeated for N cycles, then the increase in length on the Nth cycle, r_0, is given by:

$$\sqrt{r_0} = \frac{r_{p_0}}{1 - 2\alpha} \left[(1 - \alpha) - \alpha \sqrt{1 + (1 - 2\alpha) \frac{r_N}{r_{p_0}}} \right] \tag{65}$$

Initially $\delta^* < \delta_c^*$, but as cycling proceeds, a critical condition is possible since δ^* increases. Using Eq. (57), this equation may be written as:

$$\left(\frac{K_{IC}}{K_I}\right)^2 = 2\sqrt{\frac{r_N}{r_{po}}} - \alpha \frac{r_N}{r_{po}}$$

where:

$$\delta_c^* = \frac{K_{IC}^2}{\sigma_c E}$$

This process is, of course, closely analogous to the incubation time analysis and a useful approximation to N_I, the number of cycles to crack initiation, is given by assuming that $r_0 = dr_N/dN$, giving:

$$N_I = \frac{2}{\alpha}\left\{1 + \left(1 - \alpha \frac{K_{IC}^2}{K_I^2}\right)^{-1/2} + \ln\left(1 - \alpha \frac{K_{IC}^2}{K_I^2}\right)^{1/2}\right\} \qquad (66)$$

An incubation fatigue limit is given when:

$$\frac{K_I}{K_{IC}} = (\alpha)^{1/2}$$

If cycling continues after this condition, then the condition of $\delta^* = \delta_c^*$ can be maintained if the crack advances r_0 on each cycle, i.e. $da/dN \equiv r_0$, so we may write a growth law of the form:

$$\frac{da}{dN} = \frac{r_c}{(1-\alpha)^2}\left[\frac{K_I^2}{K_{IC}^2} - \alpha\right] \qquad (67)$$

where:

$$r_c = \frac{\pi}{8}\frac{K_{IC}^2}{\sigma_c^2}$$

The form of this relationship in fact is closely modelled by the Paris law over much of its range for $\alpha < 0.2$. A propagation life may also be computed in increasing K_I to K_{IC} giving the number of cycles to failure:

$$N_L = \frac{a_0}{r_c}\left\{(1-\alpha)^2 \frac{K_{IC}^2}{K_I^2} \ln\left[\frac{(1-\alpha)}{(K_I^2/K_{IC}^2) - \alpha}\right]\right\} \qquad (68)$$

The similarities of form with the time equations in Section 4 are apparent. The rate dependence of K_{IC} is ignored here, although it may be included in the analysis.

A complete description of this analysis is given in[43], together with some experimental comparisons.

7. Impact Testing

A test based on subjecting a notched sample to some form of high speed loading has many attractions as a critical test of a material. The combination of the high strain rate and the notch will often induce brittle failure where slower and unnotched tests do not. If materials perform well under these extreme circumstances, then they are frequently judged to be acceptable for demanding practical applications.

Measuring loads in high speed tests is difficult because of the short times involved and the presence of transient effects from stress waves. The more simple method which has been developed in both the Izod and Charpy tests is to break the specimen with a pendulum and measure the energy absorbed. Some form of scaling parameter from the specimen dimensions is then applied and an impact number derived. The practical utility of these numbers is beyond question but they have many, well known, drawbacks. In particular, the numbers are geometry dependent and do not agree, for example, between the Izod and Charpy tests.

Using fracture mechanics is an obvious method to effect improvement but the difficulties in making load measurements are substantial. A useful alternative is to use fracture mechanics to analyse the results of the energy measuring test and some success has been achieved with this[44-46]. Equation (9) gives the basic relationship between G_{IC} and the energy stored in the specimen, U_1, so that if it is assumed that this is that measured in the test, then by using Eqs. (23), (24) and (25), we have:

$$U_1 = G_{IC} B D \phi$$

where:

$$\phi = \frac{C}{dC/d(a/D)} = \frac{\int Y^2 x \, dx + B C_0/2}{Y^2 x}, \quad (x = a/D)$$

If ϕ is known as a function of a/D, then specimens of various crack lengths may be broken and U_1 determined. A graph of U_1 versus $BD\phi$ should be linear with a slope of G_{IC}. The ϕ functions have been determined[46] for both the Charpy and Izod geometries, which are shown in Fig. 32. That for the Charpy can be computed from published Y^2 values. The Izod should be derivable from this but such a derivation

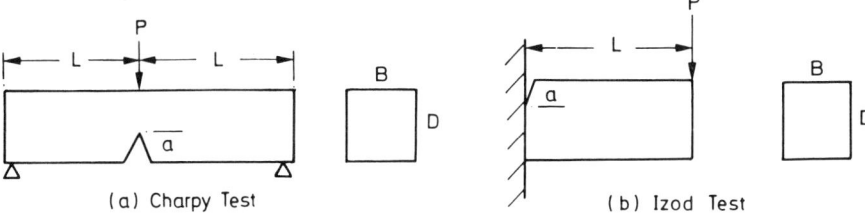

Fig. 32. Impact test geometries

assumes a rigid support which is far from true. The values given in[46] are from experimental determination of C and, in fact, C is higher than expected because of rotation at the clamp. Figure 33 shows ϕ functions for both tests for $(2\,L/D) = 4$, indicating higher Izod values.

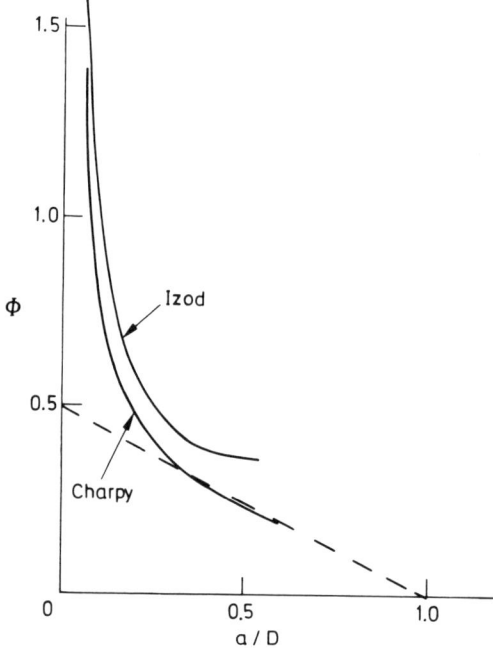

Fig. 33. ϕ function for Izod and Charpy tests for $2\,L/D = 4$ (from[46])

The effects of stress wave are not, of course, eliminated by taking energy measurements and they appear in these tests as kinetic energy errors. When the pendulum strikes the specimen, there is a rigid body impact and the specimen is accelerated away as in an unsupported collision. For a perfectly elastic impact, the energy imparted to the specimen is:

$$U_k = 2\,m\,V^2$$

where m = the mass of the specimen, and V = the velocity of impact. In fact, measurements on several polymers indicate a figure of 1.5 rather than 2 because of non-elastic effects[47].

After the first blow, this kinetic energy is converted to strain energy as the specimen bends and a further blow occurs, imparting another increment of energy. Thus, the test proceeds by a series of blows, each imparting an increment of energy and therefore it is not possible to determine the energy more accurately than one of these increments. For very brittle materials, the energy absorption:

$$G_{IC}\,B\,D\,\phi$$

is low and care must be taken that there are sufficient increments to give a sensible accuracy. A factor:

$$f = \frac{G_{IC} B D \phi}{2 \rho B D 2 L V^2} = \frac{G_{IC} \phi}{4 \rho L V^2}$$

should be considered and the various parameters adjusted to ensure that f is at least two. For these low toughness materials, however, there is still considerable scatter and the best results can only be obtained by testing a wide range of crack lengths and the drawing of a lower bound line through the origin[48]. For most materials, however, the pronounced scatter is not present, probably because of the damping effect of the higher toughness, and the energies are in error by approximately one increment. Figure 34 shows some data for polycarbonate at three temperatures showing good linearity with a common intercept but different slopes for each temperature. These lines are typical of reasonably high toughness polymers requiring about ten specimens and indicating minimum energies greater than twice U_k.

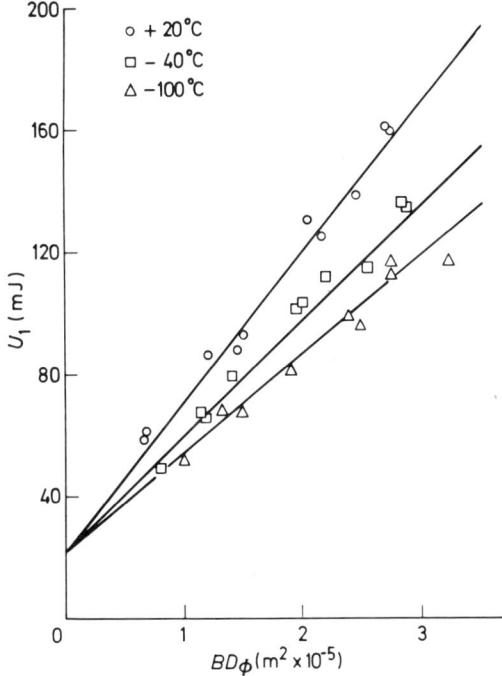

Fig. 34. Sharp notch data for polycarbonate at three temperatures (from[49])
[Reproduced from Plati, E., Williams, J. G.: Polymer 16, 915 (1975).]

For a comparison of the Charpy and Izod geometries, tests must be performed using the appropriate ϕ curves but U_k for each specimen must be determined by impacting an unsupported specimen and then subtracting from U_1. Figure 35 shows some data for polyethylene, indicating a very close agreement between the two methods using this analysis[46]. A list of several polymers tested using both methods

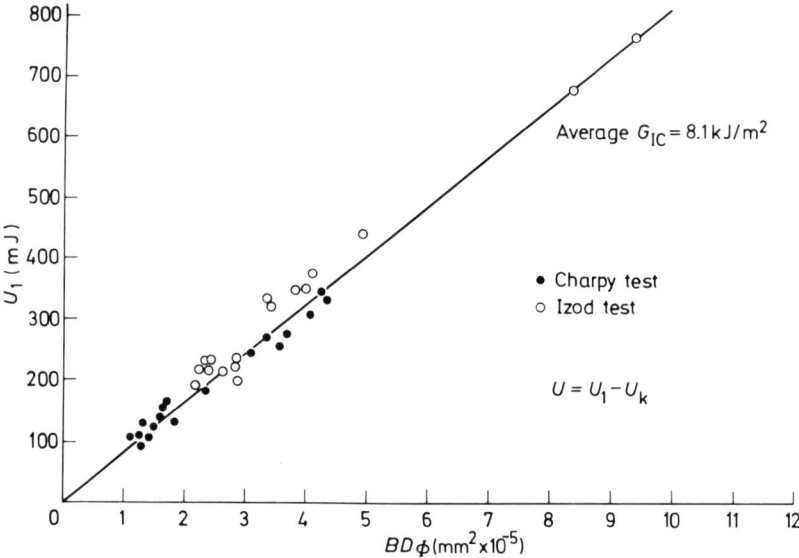

Fig. 35. Charpy and Izod impact data for polyethylene — energy to fracture as a function of $BD\phi$[46)]
[Reproduced from Plati, E., Williams, J. G.: Polym. Engn. & Sci. *15* (6), 470 (1975).]

is given in Table 1, confirming the good agreement. The values also indicate the sort of numbers to be expected. The values for high impact polystyrene (HIPS) and ABS are from ductile failures and not strictly comparable.

Table 1. Comparison of izod and charpy values[46)]

Material	G_c (kJ/m^2) Charpy	G_c (kJ/m^2) Izod
Polystyrene (GPPS)	0.83	0.83
PMMA	1.28	1.38
PVC (Darvic 110)	1.42	1.38
Nylon 66	5.30	5.00
Polycarbonate[a)]	4.85	4.83
PE (medium density)[b)]	8.10	8.40
PE (high density)[c)]	3.40	3.10
PE (low density)	34.70	34.40
PVC (modified)	10.05	10.00
HIPS	15.80 (J_c)	14.00 (J_c)
ABS (Lustran 244)	49.00 (J_c)	47.00 (J_c)

[a)] Specimens cut in the extrusion direction.
[b)] Density = 0.940; MI = 0.2.
[c)] Density = 0.960; MI = 7.5.
[Reproduced from Plati, E., and Williams, J. G.: Polym. Engn. & Sci. *15*, (6), 470 (1975).]

Applications of Linear Fracture Mechanics

For some of the tougher materials where r_p becomes large, there are errors in computing ϕ identical with those discussed in Section 3.2. Figure 36 shows data on a polyethylene showing curvature in the U_1 versus $BD\phi$ plot which is removed by the addition of a constant length (in this case, 0.5 mm) to the crack lengths[47]. For more extensive plasticity effects, such corrections are not valid and the method cannot be considered suitable for the analysis of ductile failures.

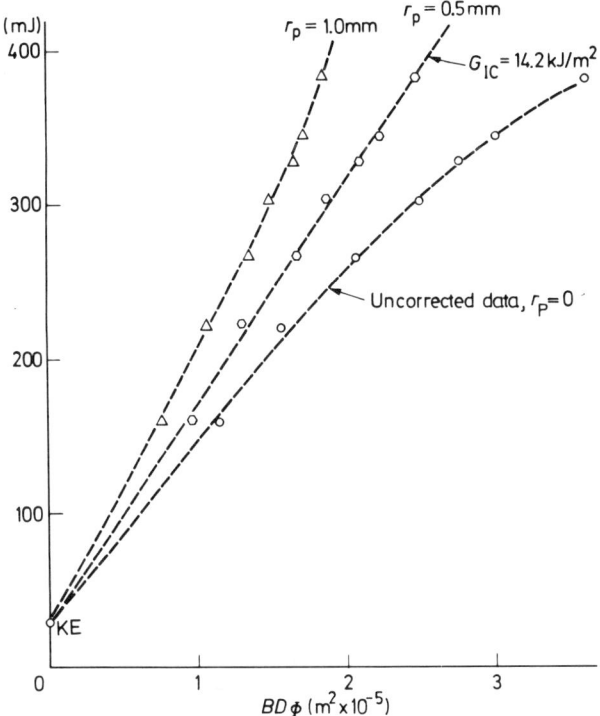

Fig. 36. Effect of adding a constant correction to the crack length[47] — polyethylene [Reproduced from Williams, J. G., Birch, M. W.: Proc. 4th Int. Conf. on Fracture, University of Waterloo, Canada, *1*, Part IV, p. 501, 1975.]

In several polymers, it is possible to induce brittle failures in impact with quite blunt notches and it is possible to give an apparent G_{IC}, G_B for these cases. Rigorously, of course, the use of K_I is not valid for anything other than a sharp notch. However, if K_{IC} is interpreted as a critical stress at a critical distance C we may write[46]a:

$$G_B = G_{IC} \frac{(1 + \rho/2\,C)^3}{(1 + \rho/C)^2} \tag{69}$$

---

a C is interpreted as r_p in[46] which is now considered questionable.

For $\rho \gg C$, this becomes:

$$G_B = G_{IC}\left(\frac{1}{2} + \frac{\rho}{8C}\right) \tag{70}$$

so that a graph of G_B versus ρ should be linear at large ρ with an intercept of $G_{IC}/2$. Figure 37 shows some typical data indicating a reasonable fit to this relationship. The C values are of the order of 50 µm for most polymers.

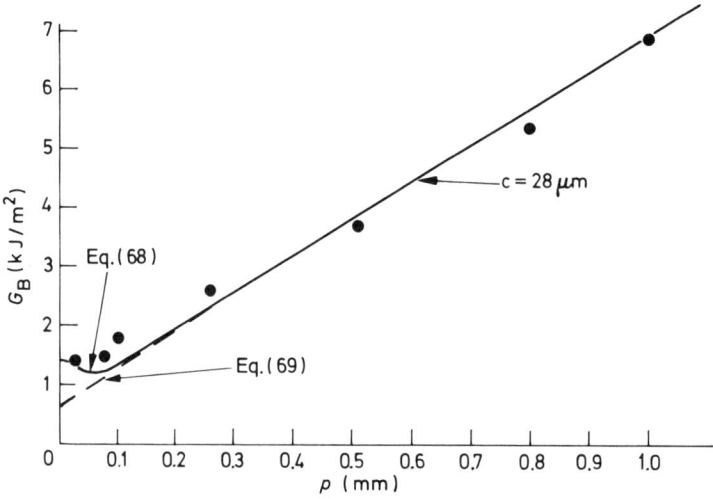

Fig. 37. Blunt notch impact data[46] for a PVC

8. Closure

The results presented here are the author's view of what is a useful framework in which to consider brittle fracture in polymers. Hopefully, it is reasonably self-contained so that those unfamiliar with the field will gain sufficient insight to see if the subject is worth pursuing further for their own purposes. Many details have, of necessity, been omitted but are given in the references. Some topics, notably the failure times and the craze growth equations, have not been published before and hence they are given in more detail.

The topics covered are reasonably wide so that most practical areas are given some attention, but our understanding at this time is not uniform. Environmental effects, particularly, have been studied in insufficient detail to give much confidence. Overall, however, the subject has reached a sufficient level of development to have some utility.

Acknowledgements. Much of this review summarises work performed in the Mechanical Engineering Department, Imperial College, London, over the past ten years or so. As the references indicate, many people have contributed and the author wishes to acknowledge his debt to them, and to the Science Research Council and BP Chemicals Limited who funded most of the work.

9. References

[1] Knauss, W. G.: Applied Mechanics Reviews 26, 1 (1973)
[2] Gurney, C., Hunt, J.: Proc. R. Soc. A299 (1967)
[3] Griffith, A. A.: Phil. Trans. R. Soc. A221, 163 (1920)
[4] Rice, J. R., in: Fracture – an advanced treatise. Liebowitz, H. (ed.). New York/London: Academic Press 1963, Vol. II, Chapter 3
[5] Williams, M. L.: J. Appl. Mech. 24, (1) (1957)
[6] Sih, G. C. (ed.): Methods of analysis and solutions of crack problems. Leyden: Nordhoff International Publishing 1973, Vol. 1
[7] Williams, J. G.: Stress analysis of polymers. London: Longmans 1973
[8] Marshall, G. P., Coutts, L. H., Williams, J. G.: J. Mat. Sci. 9, 1409 (1974)
[9] Tada, H., Paris, P. C., Irwin, G. R.: Stress analysis of cracks handbook. Del Research Corporation 1973
[10] Brown, W. F., Srawley, J. F.: ASTM, STP 410 (1966)
[11] Rooke, D. P., Cartwright, D. J.: Stress intensity factors. London: HMSO 1976
[12] Dugdale, D. S.: J. Mech. Phys. Solids 8, 100 (1960)
[13] Barenblatt, G. I.: Adv. Appl. Mechs. 7, 56 (1962)
[14] Morgan, G. P., Ward, I. M.: Polymer 18, 87 (1977)
[15] Ferguson, R. J., Marshall, G. P., Williams, J. G.: Polymer 14, 451 (1973)
[16] Irwin, G. R., Kies, J. A., Smith, H. L.: Proc. ASTM 58, 640 (1958)
[17] Parvin, M., Williams, J. G.: Int. J. Fracture 11, (6), 963 (1975)
[18] Kambour, R. P., Miller, S.: General Electric Company, CRD Report No. 77 CRD 009, March (1977)
[19] Knauss, W. G.: Int. J. Fract. Mech. 6, 7 (1970)
[20] Wnuk, M. P., Knauss, W. G.: Int. J. Solids Structures 6, 995 (1970)
[21] Kostrov, B. V., Nikitin, L. V.: Archiwum Mechaniki Stosowanej 22, English Version (1970), p. 749
[22] Schapery, R. A.: Int. J. Fracture 11, 141 (1975)
[23] Schapery, R. A.: Int. J. Fracture 11, 369 (1975)
[24] Schapery, R. A.: Int. J. Fracture 11, 549 (1975)
[25] Williams, J. G.: Int. J. Fract. Mech. 8, (4), 393 (1972)
[26] Gledhill, R. A., Kinloch, A. J.: Polymer 17, 727 (1976)
[27] Parvin, M., Williams, J. G.: J. Mat. Sci. 10, 1883 (1975)
[28] Mai, Y. M., Atkins, A. G.: J. Mat. Sci. 11, 677 (1976)
[29] Broutman, L. J., McGarry, F. J.: J. Appl. Polym. Sci. 9, 589 (1965)
[30] Williams, D. P., Evans, A. G.: J. Testing & Evaluation 1, (4), 264 (1973)
[31] Mindess, S., Nadeau, J. S., Barrett, J. D.: Wood Sci. 8, (1), 389 (1975)
[32] Williams, J. G., Marshall, G. P.: Proc. R. Soc. A342, 55 (1975)
[33] McCrum, N. G., Read, B. E., Williams, G.: Anelastic and dielectric effects in polymeric solids. New York: Wiley and Sons 1967
[34] Carslaw, H. S., Jaeger, J. G.: Conduction of heat in solids. 2nd edit. Oxford University Press 1959
[35] Andrews, E. H., Levy, G. M., Willis, J.: J. Mat. Sci. 8, 1000 (1973)
[36] Marshall, G. P., Culver, L. E., Williams, J. G.: Proc. R. Soc. A319, 165 (1970)
[37] El-Hakeem, H. M.: Ph. D. Thesis, University of London 1975
[38] Graham, I. D., Williams, J. G., Zichy, E. L.: Polymer 17, 439 (1976)
[39] Arad, S., Radon, J. C., Culver, L. E.: Polym. Eng. Sci. 12, 193 (1972)
[40] Hertzberg, R. W., Nordberg, H., Manson, J. A.: J. Mat. Sci. 5, 521 (1970)
[41] Mukherjee, B., Burns, D. J.: Exp. Mechs. 11, 433 (1971)
[42] Manson, J. A., Hertzberg, R. W.: Fatigue failure in polymers. CRC Critical Reviews in Macromolecular Science, August 1973
[43] Williams, J. G.: J. Mat. Sci., 12, 2525 (1977)
[44] Marshall, G. P., Turner, C. E., Williams, J. G.: J. Mat. Sci. 8, 949 (1973)
[45] Brown, H. R.: J. Mat. Sci. 8, 941 (1973)

[46] Plati, E., Williams, J. G.: Polym. Eng. Sci. *15*, (6), 470 (1975)
[47] Williams, J. G., Birch, M. W.: Proc. 4th Int. Conf. on Fracture, University of Waterloo, Canada, *1*, Part IV, 501 (1977)
[48] Newmann, L. V., Williams, J. G.: To be published (1977)
[49] Plati, E., Williams, J. G.: Polymer *16*, 915 (1975)

Received October 17, 1977
J. D. Ferry (editor)

Fracture and Failure of Multiphase Polymers and Polymer Composites

Clive B. Bucknall

Department of Materials, Cranfield Institute of Technology, Bedford MK 43 OAL, England

Table of Contents

List of Symbols 122

I. Introduction 123

II. Modulus 124

III. Yielding 125
    A. Mechanisms of Yielding 125
    B. Yielding of Rubber-Toughened Plastics 126
    C. Yielding of Bead-Filled Plastics 123
    D. Kinetics of Yielding 130

IV. Fracture Mechanics 135
    A. Voids 135
    B. Rigid Filler Particles 136
    C. Rubber Particles 138
        1. Linear Elastic Fracture Mechanics 138
        2. Ductile Fracture Mechanics 141

V. Impact Tests 143

VI. Concluding Remarks 146

VII. References 146

List of Symbols

a	Crack length
$\dot{a}$	Crack speed
A	Eyring pre-exponential factor
A_e	Effective cross-sectional area of specimen
A_o	Total cross-sectional area of specimen
ΔA	Area strain
B	Specimen thickness in impact test
C	Specimen compliance
$C(t,T)$ $D(t,T)$	Parameters in Oxborough and Bowden's criterion for crazing
E	Young's modulus
G_{IC}	Fracture surface energy in opening mode (Mode I)
G_{IIC}	Fracture surface energy in-plane shear mode (Mode II)
$G_{I,IIC}$	Fracture surface energy for mixture of Modes I and II
$G_{IC}(\phi)$	Fracture surface energy of composite
$G_{IC}(O)$	Fracture surface energy of matrix
h	Length of crack tip
ΔH	Activation enthalpy
I	Impact energy (impact strength)
I_{ke}	Kinetic energy contribution to impact strength
J_I	Path independent energy line integral around a crack tip
J_{IC}	Critical value of J_I at crack initiation
k	Boltzmann's constant
K_I	Stress intensity factor at crack tip
K_{IC}	Fracture toughness
$K_{IC}(\epsilon)$	Plane strain fracture toughness
$K_{IC}(\sigma)$	Plane stress fracture toughness
L	Plastic zone length
r_y	Radius of yield zone
R	Energy absorbed per unit area of ligament in impact test
t	Time
T	Temperature
T_g	Glass transition temperature
v	Activation volume
ΔV	Volume strain
$\Delta V(O)$	Volume strain at zero time under load
W	Specimen width
Z	Specimen compliance factor $C/dC/d\left(\frac{a}{w}\right)$
γ	Stress concentration factor
δ	Crack opening displacement (COD)
δ_c	Critical COD
ϵ_c	Strain at break
ϵ_1	Major principal tensile strain
$\dot{\epsilon}$	Strain rate
ϕ	Volume fraction of inclusions or voids
μ_m	Pressure coefficient in modified von Mises yield criteria
ν	Poisson's ratio
σ	Stress
$\sigma_1, \sigma_2, \sigma_3$	Principal stresses
σ_c	Critical stress at fracture
σ_y	Yield stress
σ_{yo}	Yield stress of matrix
σ_{yc}	Yield stress of composite
τ_o	Constant in modified von Mises criterion

I. Introduction

The term 'polymer composite', in its broadest sense, may be applied to a wide range of materials based on plastics or rubbers. The second phase may consist of reinforcing fibres, rigid filler particles, or voids, all of which affect fracture behaviour. However, there is little to be gained from comparing the fracture resistance of such a diverse group of materials, and the present chapter will therefore concentrate upon one broad class of polymer composites, namely, those consisting of a rigid plastics matrix containing a particulate filler. The fillers to be considered include not only glass beads and dispersions of inorganic material, but also rubber particles and voids. Whilst these fillers vary widely in rigidity, they show certain similarities in their effects upon fracture behaviour.

Perhaps the most striking effects caused by particulate fillers are those affecting the ductile-brittle transition. Ductile polymers may be embrittled by adding glass beads, especially if the beads have been treated to improve adhesion to the matrix. On the other hand, brittle polymers may be made ductile by adding rubber particles, particularly when the rubber has been grafted to improve adhesion. The mechanisms responsible for these effects are discussed in Section III below, under the heading of yielding. Changes in yielding behaviour are obviously of direct relevance to fracture resistance.

Changes in modulus also affect fracture behaviour. The relationship between the two is most clearly seen in the case of brittle fracture. For a wide, thick plate containing an edge crack of length a, the critical applied stress at fracture σ_c is related to the Young's modulus E, the Poisson's ratio ν, and the fracture surface energy G_{IC} of the material by the Griffith equation:

$$\sigma_c^2 = \frac{E\, G_{IC}}{\pi a(1-\nu^2)} \tag{1}$$

Particulate fillers alter the modulus, and thereby affect the strain energy release rate of the material under a given applied stress. Both ν and G_{IC} are also affected by the presence of a second phase, and in the absence of an introduced crack or notch the filler particles may determine the magnitude of a, by controlling the size of the largest flaws in the material.

Linear elastic fracture mechanics (LEFM), which has grown out of the work of Griffith, provides the most satisfactory available basis for the discussion of polymer composites and multiphase polymers, since it enables each of the factors contributing to fracture to be considered separately. Results obtained from fracture mechanics analyses have thrown considerable light upon the behaviour of pure polymers under tensile, impact, and fatigue loading. At present, there is only a limited amount of published information on the relationships between fracture toughness and composition in polymer composites, but a coherent picture is beginning to emerge as the subject develops. Section IV reviews progress to date, and considers current work in the field of ductile fracture mechanics, especially on rubber-toughened plastics. Extensive yielding usually precedes fracture in these materials, even in the presence of a sharp crack, so that LEFM techniques are unsuitable.

II. Modulus

Equations relating modulus to composition in isotropic composites are discussed in recent reviews by Manson and Sperling[1] and by Crowson and Arridge[2]. More than a dozen equations have been proposed. Tests on polymer composites show that none of the available equations satisfactorily predicts the moduli of all types of composite at all concentrations of rigid particulate filler; the best that can be done with any certainty is to define upper and lower bounds on the moduli.

An additional problem arises in attempting to predict the moduli of rubber-modified plastics: the rubber particles are themselves composite in structure. A typical high-impact polystyrene (HIPS) might contain only 6 vol.% of polybutadiene, whereas the volume fraction of rubber particles is between 20 and 30%. The additional volume comes from small sub-inclusions of polystyrene (PS) embedded in the rubber. The problem of relating modulus to structure in rubber-toughened plastics is discussed in a recent review[3].

Unfortunately, the distinction between nominal rubber content and rubber phase volume (*i.e.* volume fraction of rubber particles, including sub-inclusions) has not generally been recognised in the past, so that much of the published information on the relationship between moduli and composition is of little value. The rubber phase volume is considerably higher than the nominal rubber content in most rubber-toughened plastics, especially HIPS[4] and ABS. Only in the case of melt-blended polymers are the two quantities likely to be equal. In blends of polypropylene with ethylene-propylene rubber, for example, the rubber particles show no sign of sub-inclusions[5].

Additional difficulties arise in discussing the moduli of structural foams. Quite large variations in density are observed both through the thickness and across the width of a typical structural foam moulding. Consequently, it is difficult even to define a modulus. Average effective moduli may be measured in tension or in flexure, and correlated with average density, but the limitations of these correlations must be recognised[6]. There are obvious difficulties in applying fracture mechanics to materials in which structure and properties vary on a macroscopic scale in this way.

If the addition of a second phase simply alters the modulus of a brittle polymer, without affecting the intrinsic flaw size a or the fracture surface energy G_{IC}, then Eq. (1) predicts that the fracture stress σ_c will proportional to $E^{1/2}$. It follows that the strain at break ϵ_c will be proportional to $E^{-1/2}$, so that the work to break in a tensile test remains constant, independent of composition. Quite substantial changes in ultimate properties could result from the observed changes in modulus. In practice, both a and G_{IC} are affected by composition in most classes of composite.

III. Yielding

A. Mechanisms of Yielding

Two mechanisms of yielding may be distinguished in polymers: shear yielding, which occurs essentially at constant volume, and crazing, which is a dilatation process. There are important differences between the two mechanisms, both in their dependence upon stress and temperature, and in their effects on fracture behaviour.

Shear yielding in polymers has much in common with ductility in metals. In polymers, the yielding may be localised into shear bands, which are regions of high shear strain less than 1 μm in thickness; or the yield zones may be much more diffuse[7-11]. Under a general state of stress, defined by the three principal stresses σ_1, σ_2 and σ_3, the condition for yielding is given by a modified von Mises criterion[12]:

$$\frac{1}{3}\left[(\sigma_1-\sigma_2)^2 + (\sigma_2-\sigma_3)^2 + (\sigma_3-\sigma_1)^2\right]^{1/2} \geqslant \tau_o - \mu_m(\sigma_1 + \sigma_2 + \sigma_3) \qquad (2)$$

where τ_o and μ_m are materials constants. The criterion emphasises the importance of shear stresses in causing yielding. In its original form, with $\mu_m = 0$, the von Mises criterion expressly requires the presence of a shear stress to cause yielding. In its modified form, which takes account of the observed pressure dependence of yield stress in polymers, the criterion could be interpreted as predicting yielding under triaxial tension. In practice, of course, the problem does not arise, as the material always fractures before coming near the predicted yield point.

Crazing is a more localised form of yielding, and at the same time the first stage of the fracture process. A detailed discussion of the subject is to be found elsewhere in this book, and in recent reviews[3,13,14]. There is no generally accepted criterion for craze formation. The most satisfactory proposal appears to be the following, which is due to Oxborough and Bowden[15]:

$$\epsilon_1 \geqslant \frac{1}{E}\left[C(t,T) + \frac{D(t,T)}{(\sigma_1+\sigma_2+\sigma_3)}\right] \qquad (3)$$

where $C(t, T)$ and $D(t, T)$ are functions of time and temperature characteristic of the material. The criterion predicts that crazing will occur when the largest tensile strain ϵ_1 reaches a critical value.

Because of the difference in form between Eqs. (2) and (3), the mechanisms of deformation and fracture change with the state of stress. For example, polystyrene yields by shear band formation under compression, but crazes and fractures in a brittle manner under tensile loading. Changes in failure mechanism with state of stress are especially important in particulate composites, since the second phase can alter the local state of stress in the surrounding matrix.

B. Yielding of Rubber-Toughened Plastics

Yielding of rubber-toughened plastics in tension is usually accompanied by a dense whitening known as stress-whitening. It has long been known that this phenomenon is caused by multiple crazing in the matrix polymer[3, 16, 17]. Owing to their very low shear moduli, the rubber particles act as stress concentrators, transferring load to the matrix, and initiating crazes in large numbers throughout the material. By contrast with polystyrene, HIPS is able not only to yield, but also to reach elongations as high as 70%.

Ultimate elongations tend to be lower in ABS polymers, which consist of a styrene-acrylonitrile copolymer matrix containing polybutadiene particles. In addition to crazing, tensile specimens of ABS usually begin to neck immediately after yielding. The resulting increase in stress produces more intense whitening and a further decrease in cross-sectional area within the neck, leading to fracture in the neck at a total elongation of less than 20%. The low ultimate elongation simply reflects the localisation of strain within the neck region. Since crazing cannot cause a reduction in the cross-sectional area of the specimen, the formation of a neck in ABS indicates that shear yielding is taking place. It is clear that the stress concentrations produced by the rubber particles induce shear band formation as well as crazing. Both mechanisms must be considered in any complete description of yielding in polymer composites.

Bucknall and co-workers have developed a quantitative method for monitoring crazing and shear yielding independently in rubber-toughened plastics. The method involves the simultaneous measurement of axial and lateral strains during creep under uniaxial tension. Provided that the strain remains homogeneous throughout the gauge portion of the specimen, the data can be used to calculate the volume strain ΔV and the area strain $-\Delta A$ (decrease in cross-sectional area) as functions of time under load. The volume strain is divided into two components, an instantaneous elastic response $\Delta V(O)$, which is calculated from the initial strain readings immediately after loading, and a time-dependent volume change, which is taken as a measure of crazing[3], since shear deformation takes place at constant volume. Similarly, the area strain is divided into an instantaneous elastic response and a time-dependent area change. Since crazing makes no contribution to ΔA, the time-dependent area strain is a measure of shear deformation. The method has been applied to HIPS[17], HIPS/PPO blends[18, 19], ABS[20], ASA (acrylonitrile-styrene-acrylate polymer[3, 21], toughened epoxy resins[22], toughened PVC[3, 23], and toughened polypropylene[23].

This technique reveals significant differences between materials in their response to tensile stress, as illustrated in Fig. 1. The creep of ASA polymer, which is dominated by crazing, begins slowly, and then accelerates. On the other hand, the creep of polypropylene, which is dominated by shear yielding, begins relatively rapidly, and then decelerates. The difference in mechanism is evident from Fig. 1, but is brought out even more clearly in Fig. 2, in which volume strain is plotted against axial strain. When the data are plotted in this way, a slope of unity indicates that the material is deforming entirely by crazing, whereas a slope of zero indicates that the material is deforming entirely by shear mechanisms. The two polymers chosen for the illustration approximate to these extremes.

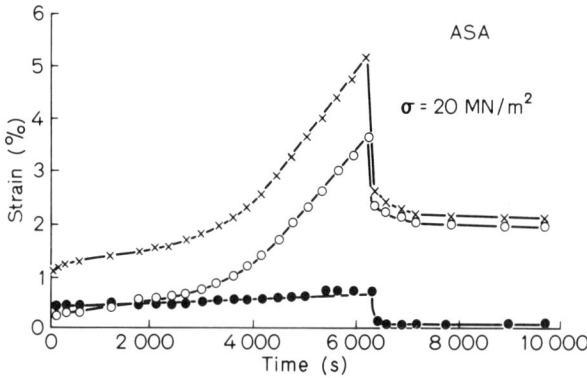

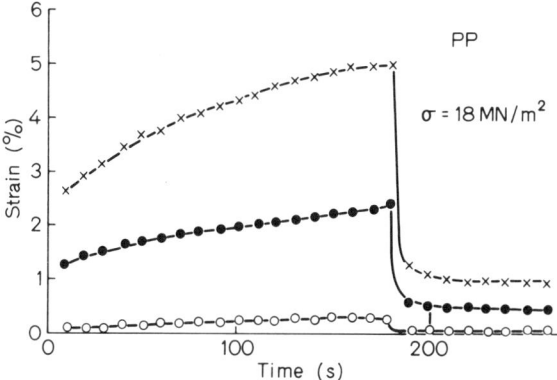

Fig. 1. Tensile creep and recovery of rubber-modified plastics, showing axial strain (x), lateral contraction (●), and volume strain (○) for ASA polymer and polypropylene copolymer at 20 °C

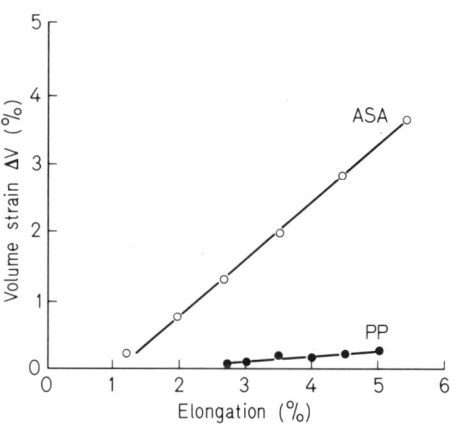

Fig. 2. Relationship between volume strain and axial strain during creep of ASA polymer and polypropylene copolymer at 20 °C., showing difference in deformation mechanism

Composition is not the only factor affecting the mechanism of deformation in rubber-modified plastics. Strain, strain rate, stress configuration, temperature, and molecular orientation all influence the balance between crazing and shear yielding. As already noted, ABS necks at 20 °C, whereas HIPS does not. On raising the temperature to 60 °C, however, HIPS also necks in a tensile test, showing the same combination of crazing and shear yielding as ABS at lower temperatures. This observation reflects differences in the kinetics of the two mechanisms. Injection-moulded HIPS bars will neck at 20 °C if the orientation in the gauge portion is high enough. The reason is that orientation parallel to the applied tensile stress inhibits crazing[24].

The contribution of crazing to tensile deformation increases with stress, and therefore with strain rate. This point is illustrated in Fig. 3, which shows changes in mechanism with stress and strain in ABS and in toughened epoxy resin. The data are taken from creep tests in which the strain rate increased with time under load, so that the values quoted for low strains are also for low strain rate. The changes in

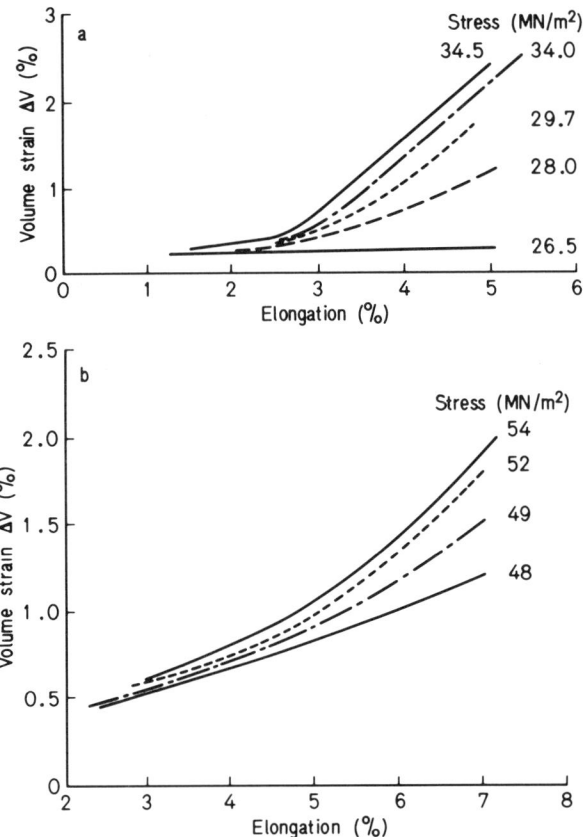

Fig. 3. Relationship between volume strain and axial strain during creep of (a) ABS polymer, and (b) epoxy resin containing CTBN rubber, at 20 °C. Note magnification of volume strain scale in (b)

mechanism shown in Fig. 3 again reflect differences in kinetics between crazing and shear yielding. One interesting consequence of the changes in mechanism is that the elongation at break of ABS can actually increase with strain rate: the contribution of shear yielding to the elongation decreases, so that the specimen reaches high strains without developing a significant neck. Using high speed cinematography to measure strains, Fenelon and Wilson[25] and Cessna[26] have studied volume changes in ABS specimens under high-velocity tensile impact loading, and shown that crazing dominates the deformation at strain rates in the region of $100 \, s^{-1}$.

The transition from shear yielding to multiple crazing in rubber-toughened plastics may be compared with the ductile-brittle transition in unmodified plastics. Both occur when an increase in strain rate or a decrease in temperature raises the shear yield stress above the stress required for craze formation. The significant difference in rubber-toughened plastics is that rubber particles change crazing from a fracture process to a yield mechanism. The multiple crazing mechanism is most effective when some shear yielding is also present. Shear bands appear to control craze growth, preventing any individual craze from becoming too large, an effect that can be attributed to the high molecular orientation within shear bands[18, 19].

C. Yielding of Bead-Filled Plastics

Rubber particles are not unique in their ability to induce tensile yielding in brittle glassy polymers. Nicolais and co-workers have shown that styrene-acrylonitrile (SAN) copolymers and a number of other glassy polymers can be made to yield by adding glass beads[27, 28]. Like rubber-toughened plastics, the bead-filled plastics stress-whiten at the yield point, owing to multiple crazing. However, unlike rubber-toughened plastics, the glass-filled materials have elongations at break that are only

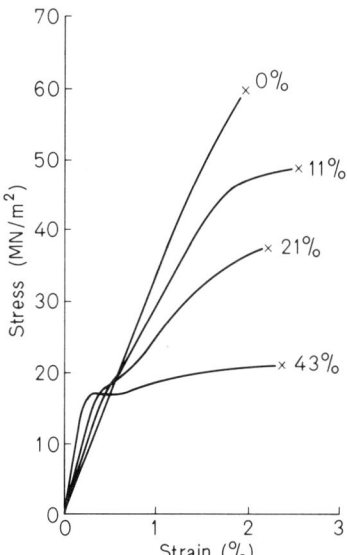

Fig. 4. Tensile stress-strain curves for SAN containing 0–43 vol.% glass beads; after Lavengood et al. (27)

slightly greater than those of the parent polymers, and the work to break is actually reduced on adding beads, so that there is no toughening effect. Clearly the beads are able to initiate crazing by transferring load to the matrix, but they appear to be unable to control craze growth effectively. Stress-strain curves for glass-filled SAN are shown in Fig. 4.

Glass beads have an equally striking effect upon the stress-strain behaviour of ABS, as shown in Fig. 5. The elongation at break increases dramatically, from 10.5% for ABS itself, to over 80% for a specimen containing 11% of glass beads[27, 28]. As

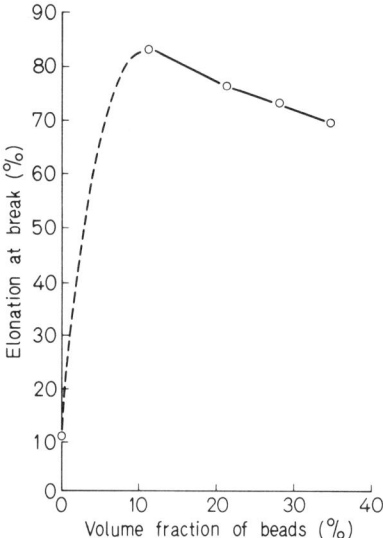

Fig. 5. Ultimate elongation of ABS/glass bead composites as a function of bead content; after Lavengood et al. (27)

in SAN, the unbonded beads produce an increased level of crazing in the surrounding polymer. There is, however, no comparable increase in the level of shear yielding, presumably because of the rigidity of the beads. The result is that necking is suppressed, a much larger volume of the specimen becomes involved in yielding, and the elongation at break increases. This is another good example of a change in mechanism of deformation in a polymer composite.

D. Kinetics of Yielding

Creep and yielding are stress- and temperature-activated processes, which in many materials, including polymers, follow the Eyring rate equation[30]:

$$\dot{\epsilon} = A \: exp \left(\frac{\gamma v \sigma - \Delta H}{kT} \right) \qquad (4)$$

where $\dot{\epsilon}$ = strain rate, γ = stress concentration factor, v = activation volume, ΔH = activation enthalpy, k = Boltzmann's constant, and A is a constant for a given material. Equation (4) may be rearranged as follows to show the relationship between yield stress σ_y and strain rate in a tensile test[31]:

$$\frac{\sigma_y}{T} = \frac{k}{\gamma v}\left(\log_e \frac{\dot{\epsilon}}{A} + \frac{\Delta H}{kT}\right) \tag{5}$$

The Eyring model assumes that deformation is made up of a series of individual steps, in which the molecules pass over energy barriers, with the aid of thermal and mechanical energy. The applied stress biases the direction in which these steps are made, producing an overall strain in the material.

Evidence for differences in activation volume and enthalpy between shear yielding and crazing has already been presented. In discussing kinetics, it is convenient to treat the two mechanisms as independent, and to calculate activation parameters for each process accordingly. It must be noted, however, that interactions do occur between crazes and shear bands under certain conditions, so that the kinetics cannot be regarded as completely independent.

The presence of a void or inclusion will obviously affect the stress concentration factor γ. Stress concentrations around an isolated spherical inclusion may be calculated using Goodier's equations[32], but this approach cannot be used for a composite containing a high proportion of particles. One answer to the problem is to obtain a numerical solution using finite element analysis. Having made the analysis, however, it is not easy to apply the results to predict yield behaviour. A much simpler approach based on the calculation of effective cross-sectional areas offers many advantages, although its limitations must be recognised. The effective area model, proposed by Ishai and Cohen[33], ignores local variations in stress within the matrix, and simply calculates an average stress borne by the matrix in those regions where yielding is taking place. According to the model, the stress concentration factor γ is the ratio of the total area A_o of the specimen to the effective area A_e of the matrix. For a material containing a volume fraction ϕ of voids, the effective area is given by:

$$\gamma = \frac{A_o}{A_e} = \left[1 - \pi\left(\frac{3\phi}{4\pi}\right)^{2/3}\right]^{-1} = \left(1 - 1 \cdot 21\,\phi^{2/3}\right)^{-1} \tag{6}$$

The effective area is calculated on the assumption that the shear bands or crazes responsible for yielding follow a minimum-area path through the matrix, and therefore pass through the centre of each void that they encounter. The relationship between A_e and ϕ is easily obtained by considering a cube having a spherical hole at its centre, and calculating the area outside the sphere in a plane through the equator, in terms of ϕ. Ishai and Cohen also considered the case of composites containing rigid filler particles that are well bonded to the matrix. They concluded that in this case the stress concentration factor should be unity, since the path of least resistance for the yield zones lies entirely within the matrix.

Despite its limitations, the effective area model has been used with some success by a number of authors. The most comprehensive test of the model is to insert the predicted value of γ into the Eyring equation, and to make measurements over a wide range of temperatures, strain rates and compositions to check the validity of the resulting equation. Combining Eqs. (4) and (6) yields:

$$kT \log_e \left(\frac{\dot{\epsilon}}{A} \right) = \frac{v\sigma}{\left(1 - 1.21 \phi^{2/3} \right)} - \Delta H \tag{7}$$

which on rearranging becomes:

$$\sigma_y = \frac{kT}{v} \left(1 - 1.21 \phi^{2/3} \right) \left(\log_e \frac{\dot{\epsilon}}{A} + \frac{\Delta H}{kT} \right) \tag{8}$$

Ishai and Cohen tested their model by measuring yield stresses in uniaxial compression over a range of strain rates for a series of epoxy resins containing 0 to 66% of voids. Their results are summarised in Fig. 6. As predicted by Eq. (8), yield stress increases with log (strain rate), and the slope changes systematically with void

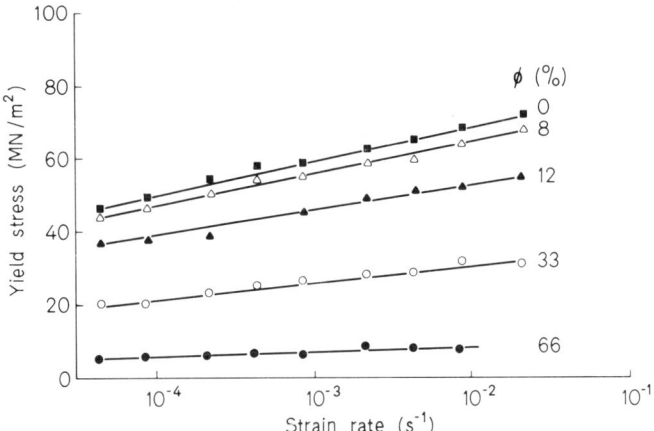

Fig. 6. Eyring plots of compressive yield stress against strain rate for epoxy resins containing 0–66 vol.% voids; after Ishai and Cohen (33)

content. Attempts to produce a master curve by plotting $\sigma_y/(1 - 1.21\phi^{2/3})$ against $\log \dot{\epsilon}$ were not entirely successful: the lines obtained had identical slopes, but did not coincide. The discrepancy appears to be due to experimental error, since the relative displacements are not systematically related to void content.

Under tensile loading, the stress concentrating effect of an unbonded spherical particle is similar to that of a void. Nicolais and co-workers have studied the tensile stress-strain behaviour of composites based on SAN, ABS, PPO, and epoxy resins[27–29, 34–36]. In each of these polymers, unbonded glass beads cause yielding accompanied by stress-whitening, which suggests that failure of the weak bond between polymer and glass is followed by crazing in the surrounding matrix. The effective area model predicts that the yield stress σ_{yc} of the composite is related to the yield stress of the matrix σ_{yo} and to the volume fraction of beads ϕ by the equation:

$$\sigma_{yc} = \sigma_{yo} (1 - 1.21\, \phi^{2/3}) \qquad (9)$$

This equation cannot be tested for SAN or epoxy resin because these polymers do not yield in tension. Tests on ABS and PPO show a reasonable agreement with Eq. (9). Nicolais and DiBenedetto found that the yield stress of bead-filled PPO composites varied in accordance with Eqs. (8) and (9) over a wide range of temperature, strain rate, and composition[36].

Rubber-modified plastics are another class of composites to which the effective area model might be applied. Unbonded rubber particles would be expected to have the same effect as unbonded glass beads or voids on the stress distribution in the surrounding matrix under conditions of tensile loading. However, very few products contain unbonded rubber particles, since the toughening effect is greatly enhanced by grafting the rubber to ensure good adhesion to the matrix[37]. A grafted rubber particle may act like a void, and carry very little load, or it may share some of the load with the matrix. Under uniaxial tensile stress, the deciding factor is the shear modulus of the rubber particle. Rubbers have very low shear moduli compared with glassy polymers, so that a pure rubber particle is unable to support a significant level of tensile stress. On the other hand, grafted rubber particles often contain a high proportion of hard sub-inclusions, which raise the shear modulus of the particle (3), so that some degree of load sharing becomes possible. Equation (9) is of course not applicable to materials in which the particles share the load with the matrix. Another point to be borne in mind when discussing rubber-modified plastics is that rubbers have relatively high bulk moduli: under triaxial tension or compression, bonded rubber particles are certainly capable of sharing the load with the matrix. The possible effects of triaxial stresses at a crack tip on the formation of a yield zone in rubber-modified plastics are discussed in Section IV.

As explained earlier, most authors quote nominal rubber contents rather than rubber phase volumes, and there is therefore very little information in the literature on the relationship between σ_{yc} and ϕ for rubber-modified plastics. A rare exception occurs in the work of Oxborough and Bowden[38], who measured yield stresses in tension and compression for a series of HIPS polymers containing composite rubber particles. Their results are presented in Fig. 7. Equation (9) underestimates the yield stresses both in tension and compression, and it must be concluded that the effective area model does not provide a satisfactory basis for correlating yield data in this class of material. Either the model itself must be modified in some way, or some allowance must be made for load sharing with the rubber particles, if the effective area approach is to be retained.

Roylance *et al.* studied yielding at temperatures between 0 and 75 °C in AMBS polymers, which are poly (styrene-acrylonitrile-methyl methacrylate) copolymers containing polybutadiene particles, and showed that Eq. (5) is obeyed over three decades of strain rate[39]. Nominal rubber contents varied between 0 and 13%, and the slopes of the Eyring curves changed with rubber content in the direction predicted by Eq. (8). As rubber phase volumes are not given, the data cannot be

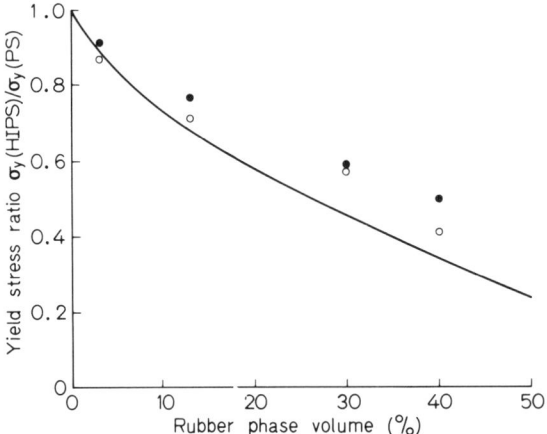

Fig. 7. Yield stress of HIPS, expressed as a fraction of the yield stress of PS, as a function of rubber phase volume. Data obtained by Oxborough and Bowden (*38*) in tensile (o) and compressive (•) tests are compared with Eq. 9 (solid line)

checked quantitatively against this equation. Whilst the rubber particles in AMBS polymers are grafted, they do not contain the large sub-inclusions that are characteristic of HIPS, and might therefore be more suitable subjects for the effective area model.

The problems of correlating yield data for rubber-toughened plastics are further illustrated by the work of Imasawa and Matsuo[40], who measured yield stresses over a wide range of temperatures and strain rates for a series of blends of PVC with

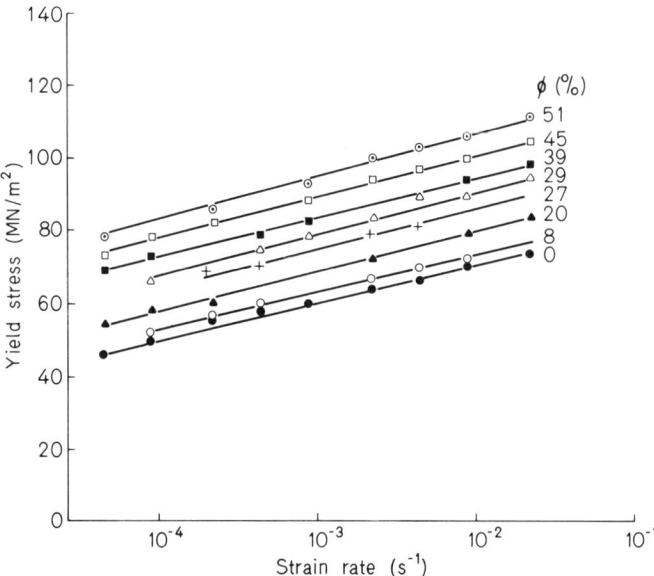

Fig. 8. Eyring plots of compressive yield stress against strain rate for epoxy resins containing 0–51 vol.% glass beads; after Ishai and Cohen (*33*)

nitrile rubbers. The results were in good agreement with the Eyring equation, and the slopes varied with rubber content. However, the change in slope with composition was far greater than that predicted by Eq. (9). The reason for this strong dependence of stress concentration factor upon rubber content was revealed by electron microscopy: the rubber formed network structures rather than discrete particles[41]. Once again, the nominal rubber content does not provide sufficient information for the prediction of mechanical properties.

The effective area model predicts that the stress concentration factor γ should be independent of composition in composites containing well-bonded rigid filler particles. This prediction is supported by the compressive yield data for silica-loaded epoxy resins presented in Fig. 8: yield stress is linear with log (strain rate) for each material, and the slopes are identical in each case. The increase of yield stress with silica content must therefore be interpreted as a decrease in the pre-exponential factor A rather than in γ. Young and Beaumont observed a similar large increase in the yield stress of silica-loaded epoxy resins, and suggested an analogy with precipitation-hardening in metals[42].

IV. Fracture Mechanics

The basic principles of this subject, including the definition of fracture toughness K_{IC}, and the difference between plane stress and plane strain fracture, are set out in the contribution of J. G. Williams.

Fracture mechanics studies have tended to concentrate upon single-phase polymers, and relatively few papers have devoted attention to the fracture toughness of polymer composites. The subject has now developed to the stage at which fracture mechanics tests can be regarded as routine measurements, and data on all types of composites are beginning to appear. The results provide a useful insight into the factors affecting strength in two-phase materials.

A. Voids

There is at present very little published information on the fracture toughness of polymeric foams. Two papers give G_{IC} values for low-density rigid polyurethane foams[43, 44], but the relationship between toughness and void content is not discussed in either case.

In a brittle polymer, G_{IC} measures the energy absorbed in forming unit area of fracture surface. Voids reduce the effective area of the fracture surface, and the effective area model predicts that the fracture surface energy $G_{IC}(\phi)$ of a foam should be related to the volume fraction ϕ of voids and the fracture surface energy $G_{IC}(0)$ of the matrix by the equation:

$$G_{IC}(\phi) = G_{IC}(0)(1 - 1 \cdot 21\phi^{2/3}) \tag{10}$$

Akhurst and Bucknall have measured the fracture toughness of HIPS structural foams over a range of densities, and obtained results[45] that are in reasonable agreement with Eq. (10).

Hobbs found that fracture toughness actually increased with void content in polycarbonate foams containing 5% short glass fibres[46]. The unfoamed polymer gave a G_{IC} of approximately 11 kJ/m^2, whereas a structural foam containing 20% voids gave a G_{IC} of 14 kJ/m^2. Hobbs suggested that the increase in fracture toughness was due to shear yielding in the cell walls. Constraints at the crack tip prevent yielding in the solid polymer, and the introduction of voids reduces the constraints, to permit a limited amount of yielding. The bulk modulus of the voids is zero, so that the hydrostatic tension component of stress at the crack tip is reduced.

B. Rigid Filler Particles

Unbonded glass beads appear to have a similar effect to voids on the constraints at a crack tip. DiBenedetto and co-workers[47, 48] measured G_{IC} in PPO composites containing 0–25% of bonded and unbonded glass beads, using double-edge-notch specimens between 1.5 and 5 mm thick. The unfilled polymer was brittle, with a smooth, glass-like fracture surface, over the whole range of thicknesses. The composites containing unbonded glass beads showed a transition in fracture behaviour with increasing thickness, from predominantly plane-stress fracture to predominantly plane strain fracture; most of the specimens showed evidence of mixed-mode failure on the fracture surface, with limited regions of plastic flow visible in the scanning electron microscope. Figure 10 shows the relationship between G_{IC} and glass content. The decrease in fracture resistance is greatest when the beads have been treated to improve adhesion to the matrix. These observations suggested that partial debonding at the glass-polymer interface relaxes the constraints at the crack tip, and permits a limited amount of shear yielding.

As shown in Fig. 9, the decrease of G_{IC} with glass content in PPO composites is approximately in agreement with Eq. (10). The beads reduce the area of polymer

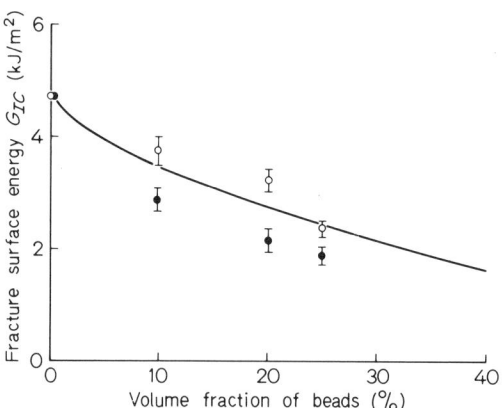

Fig. 9. Relationship between G_{IC} and filler content in PPO composites containing glass beads: (○) without coupling agent; and (●) with coupling agent. Solid line is Eq. 10. After Trachte and DiBenedetto (48)

on the fracture surface, and G_{IC} decreases in the manner predicted by the effective area model. The applicability of this model implies:

(a) that the crack follows a least-area path through the matrix;

(b) that the energy absorbed in overcoming adhesion at the glass-PPO boundary is negligible; and

(c) that the filler particles do not introduce additional deformation processes.

Condition (c) is not completely fulfilled in composites containing unbonded glass beads. Scanning electron microscopy and the higher values of G_{IC} in this type of composite show that unbonded beads promote some additional yielding. A similar effect of unbonded beads upon yielding mechanisms[36] is discussed in Section III.

Unlike PPO, epoxy resins are *toughened* by the addition of rigid filler particles. DiBenedetto found that G_{IC} increased monotonically on adding glass beads[49, 50]. Figure 10 presents results obtained using alumina trihydrate[51] and silica[42], both

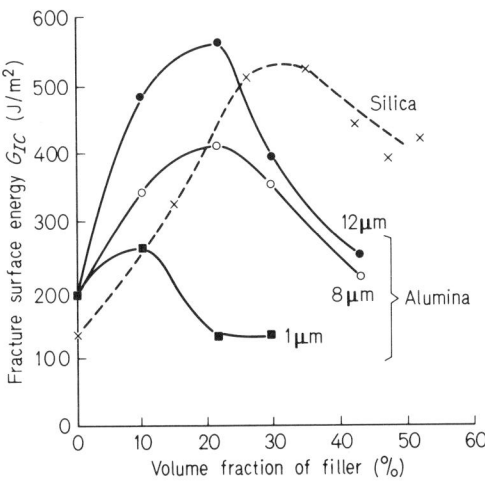

Fig. 10. Relationship between G_{IC} and filler content in epoxy resin composition containing 70 μm silica particles (*42*) and alumina trihydrate particles of differing sizes (*51*)

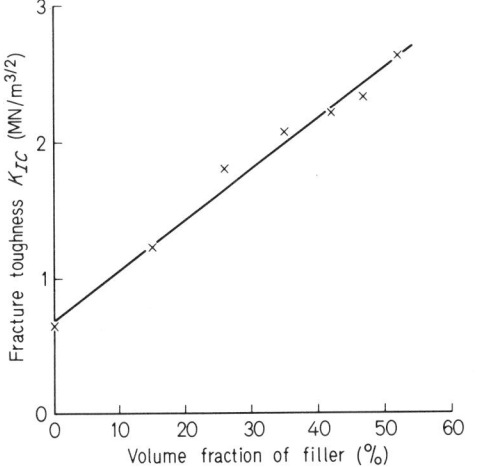

Fig. 11. Relationship between K_{IC} and filler content in epoxy resin composites containing 70 μm silica particles (*42*)

of which produce a maximum in G_{IC} at relatively low volume fractions. Although G_{IC} begins to fall, the modulus continues to increase with filler content, with the result that K_{IC} also increases continuously, as shown in Fig. 11. These increases in toughness with filler content are especially striking when PPO and epoxy resin composites are compared. The experimental data strongly suggest that filler particles initiate additional energy-absorbing deformation processes, perhaps including crazing, in the epoxy resin matrix. The factors that determine whether a rigid particulate filler will increase or decrease the fracture surface energy of a given polymer are far from clear, and deserve further study.

C. Rubber Particles

Rubber particles are added to plastics with the specific aim of increasing fracture resistance. The multiple crazing mechanism enables rubber-modified plastics to absorb large amounts of energy by yielding at the crack tip. As linear elastic fracture mechanics (LEFM) methods are invalid when there is extensive yielding in the specimen, it is necessary to employ ductile fracture criteria to characterise this type of failure. The alternative is to establish conditions under which yielding is severely restricted, so that LEFM measurements can be made. Both approaches are discussed below.

1. Linear Elastic Fracture Mechanics

In order to make valid measurements of G_{IC} and K_{IC} on normally ductile materials, it is necessary to use specimens having the correct geometry. Firstly, the specimen must be thick enough to ensure that plane strain conditions are established at the base of the notch or crack. Secondly, the specimen must be wide enough to prevent net section yielding prior to crack initiation; with narrow specimens there is a danger that the yield zone formed at the crack tip might interact with the edge of the specimen, or even produce a state of general yielding. Thirdly, the crack tip radius must be small. Each of these factors plays an important part in restricting yielding.

Owing to the difficulty of moulding thick sheets, most fracture measurements on polymers have been made on specimens less than 7 mm thick. This thickness is not sufficient to produce plane strain conditions in most tough or ductile polymers. The difference between plane stress and plane strain fracture has hitherto been widely regarded as unimportant in HIPS and other rubber-modified plastics, since crazing is itself a plane-strain process, and little attempt has been made to work with thick specimens. However, recent work by Parvin and Williams has shown that there is a very marked transition between plane stress and plane strain fracture in HIPS[52], and it is clear that this aspect of rubber-toughening will require closer attention in the future.

Parvin and Williams made fracture toughness measurements on HIPS between $-120°$ and $20\ °C$. Some of the measurements were made on 150 mm wide single edge (SEN) specimens, but the study concentrated on fracture in 6 mm thick surface notch (SN) specimens, in which the notch was machined into the broad face of the sheet. This geometry provides a greater degree of constraint than is possible with SEN specimens of the same thickness, and therefore comes closer to the ideal of plane-strain crack initiation. A possible criticism of the test is that the notch tip is too close to the opposite face of the sheet, so that some interaction could occur; the distance is only a few millimetres.

A craze-yield zone of length r_y was observed ahead of the crack in the plane stress region of the specimen. It was assumed that this yield zone extended inwards to a depth of r_y on each side of the specimen, leaving a plane strain region with a width of $(h - 2r_y)$ at the centre of the crack front, where h is the length of the crack front. In a SEN test, h is the specimen thickness. It was further assumed that the overall fracture toughness of a specimen was the sum of two terms, a plane stress term $2r_y K_{IC}(\sigma)$ and a plane strain term $(h - 2r_y)K_{IC}(\epsilon)$, where $K_{IC}(\sigma)$ and $K_{IC}(\epsilon)$ are the plane-stress and plane-strain fracture toughnesses of the material. The authors give no theoretical justification for treating K_{IC} as an additive quantity in this way[52, 53].

The results of this analysis are given in Fig. 12. Parvin and Williams made the further assumption that the plane strain fracture toughness $K_{IC}(\epsilon)$ of HIPS was equal to K_{IC} for unmodified polystyrene, i.e. that the multiple crazing mechanism is inoperative in HIPS under ideal plane-strain conditions. Values of $K_{IC}(\sigma)$ calculated on this basis reach a maximum at -90 °C., the glass transition of polybutadiene, which suggests that energy absorption at the crack tip is due largely to viscoelastic processes in the rubber phase. Although overall values of K_{IC} were substantially different for SEN and SN specimens, the differences were successfully resolved using the additive model outlined above.

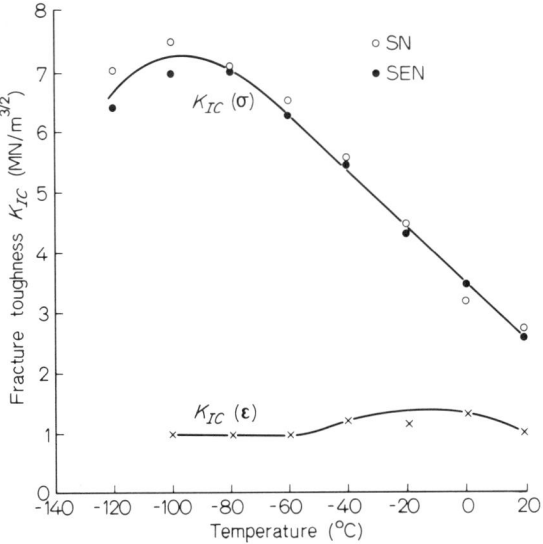

Fig. 12. Plane-strain and plane-stress components of fracture toughness of HIPS as a function of temperature; after Parvin and Williams (52)

The results presented in Fig. 12 are at first sight rather surprising, since standard tests show that the fracture resistance of HIPS and other rubber-modified plastics increases with temperature. The difference between the two types of test lies in the extent of yielding. The fracture mechanics specimens were designed to produce brittle fracture, with the minimum of yielding; crack geometry, tip sharpness, specimen width, and specimen thickness were all chosen accordingly. The resistance to the initiation of *brittle* fracture follows the trends shown in Fig. 12. On the other hand, *ductile* fracture resistance increases with temperature, as the yield stress falls. Parvin and Williams measured K_{IC} in 5 mm wide SEN specimens, and found that toughness increased with temperature above $-60\ °C$.

The form of the stress field at the tip of a sharp crack is given by the Westergaard equations. Under plane strain conditions, the three principal stresses are in the ratio $1:1:2\nu$. Poisson's ratio ν for polystyrene is 0.42, so that the hydrostatic component of stress is large compared with the deviatoric component. Parvin and Williams suggested that the low value of $K_{IC}(\epsilon)$ for HIPS was due to the inability of crazes to form in a triaxial stress field, and quoted Eq. (3) in support of this argument: the application of lateral stresses σ_2 and σ_3 reduces the tensile strain ϵ_1, and so inhibits crazing. Another factor contributing to the low value of $K_{IC}(\epsilon)$ is the ability of the rubber particles to support stress when subjected to triaxial loading. This consequence of the high bulk modulus of rubber has already been mentioned in Section III.D. Rubber-toughening depends upon the stress concentrating effect of the rubber particles, and under plane strain conditions at a crack tip the magnitude of the stress concentrations is likely to be greatly reduced. The difference between $K_{IC}(\sigma)$ and $K_{IC}(\epsilon)$ can also be explained in terms of the effect of triaxiality on the rubber particles. In the plane stress region, where the particles are under biaxial loading, the rubber is able to relax in response to the applied load, and gives a viscoelastic loss maximum at the glass transition. In the plane strain region, constraints prevent such a relaxation, and the loss factor is greatly reduced.

It is clear from this work that most of the published values of K_{IC} and G_{IC} for rubber-toughened plastics refer to plane-stress fracture. Measurements have been made on a variety of polymers using SEN, double cantilever beam (DCB) and Charpy impact specimens. Figure 13 shows the results of one such study by Kobayashi and Broutman[54], who used DCB specimens to measure G_{IC} in AMBS polymers over a range of crack speeds. The two most prominent features of the results are the rapid rise in G_{IC} with rubber content, and the fall in G_{IC} at high crack speeds. Both effects are predicted by Eq. (8): the yield stress decreases with increasing rubber phase volume, so that the size of the plastic zone at the crack tip increases; similarly, increasing crack speed (and therefore strain rate at the crack tip) increases yield stress and reduces the size of the plastic zone. Thus the yield stress is the link between rubber phase volume and fracture resistance.

For obvious reasons, fracture mechanics studies have concentrated on opening mode (Mode I) fractures, and only a few studies have been made of other fracture modes. In one such study, Bascom *et al.* performed mixed-mode fracture tests on epoxy resin adhesives, with the epoxy layer at $45°$ to the applied stress[55]. This test combines the opening Mode I with the in-plane shear Mode II. The authors calculated a mixed-mode fracture energy $G_{I,\ IIC}$ from the results, and obtained a

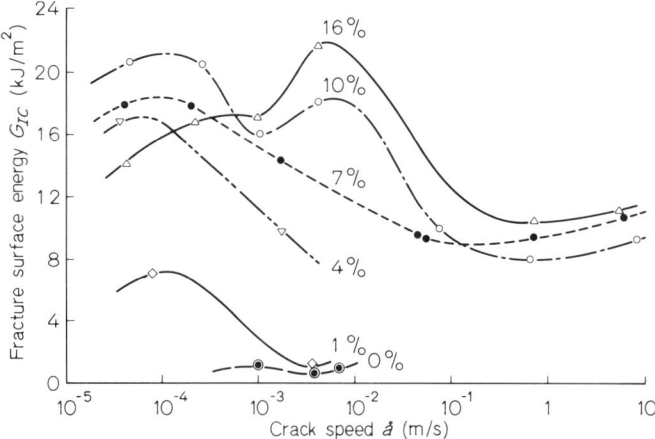

Fig. 13. Effects of crack speed and rubber content on G_{IC} of AMBS polymers; after Kobayashi and Broutman (54)

value of 140 J/m² for epoxy resin, whereas G_{IC} for the same resin was 116 J/m². Adding rubber to the resin increased G_{IC} to 3500 J/m², but reduced $G_{I,IIC}$ slightly, to 110 J/m². These results are not unexpected. Rubber particles increase fracture resistance under opening-mode conditions by initiating multiple crazing. However, this mechanism is inoperative under shear loading, and the addition of a rubber gives no improvement. The small decrease in $G_{I,IIC}$ can perhaps be related to a decrease in effective area.

2. Ductile Fracture Mechanics

The problem of defining conditions for crack propagation becomes more difficult when the material is sufficiently ductile to form a large plastic zone at the crack tip. The problem is encountered particularly in testing rubber-modified plastics: other types of particulate filler do not increase ductility to the same extent, if at all. Two ductile fracture criteria have been developed for metals, one based on crack opening displacement (COD), and the other upon the energy line integral (J-integral) around the crack tip. The COD criterion has been applied to a number of polymers, but the more recent J-integral criterion has received relatively little attention as yet.

The COD δ is the relative displacement of the two fracture surfaces at the crack tip. Critical values of COD $δ_c$ may be measured by means of a mechanical clip gauge[56] or recorded photographically. Ferguson et al.[57] used a cine camera to follow crack initiation and propagation in HIPS at 20 °C, and found that both LEFM and COD criteria were applicable to different stages of fracture. On application of load to the specimen, the crack began to extend at a fixed value of K_I, to give a value of K_{IC} for crack initiation. A stress-whitened yield zone then began to form at the crack tip. The load on the specimen continued to increase as both crack and yield zone extended, and a load maximum was observed at a fixed value of COD.

Whilst direct measurements of COD provide the best evidence for the validity of the criterion, they are inconvenient to make, and often difficult to make with any precision. Furthermore, COD values are unsuitable for use in design, which is usually based on applied loads and component geometry. For these reasons, there are great advantages in developing models that enable COD to be calculated in terms of stresses. The most widely used model is that of Dugdale[58], who considered the case of a central crack of length $2a$ in a very wide plate, with a narrow planar plastic zone of length L extending from each of the crack tips. The plastic zone length is related to the applied stress σ by:

$$\frac{a}{a+L} = \cos\left(\frac{\pi\sigma}{2\sigma_y}\right) \tag{11}$$

The COD is then given by [56]:

$$\delta = \frac{8\sigma_y a}{\pi E} \log_e \sec\left(\frac{\pi\sigma}{2\sigma_y}\right) \tag{12}$$

For small values of applied stress ($\sigma < 0.3\,\sigma_y$), the plastic zone size is small compared with the crack length, and Eq. (12) reduces to:

$$\delta = \frac{\pi \sigma^2 a}{E \sigma_y} \tag{13}$$

Under these conditions, LEFM analysis is applicable to the specimen. The relationship between COD and brittle fracture criteria can be seen by combining Eq. (13) with Eq. (1), to give:

$$G_{IC} = \sigma_y \delta_c \tag{14}$$

In practice, tests are performed on SEN or compact tension specimens rather than wide centre-notched plates, and some allowance must be made for geometrical effects, including the finite width of the test piece, the difference between edge and centre notches, and any rotations occurring at the grips. Ferguson et al. used a finite-element analysis approach to this problem in their work on HIPS[57]. More recently, Hoffman and Richmond[59] compared this method and two other solutions to the problem with experimental data on δ_c and L for HIPS at room temperature. They concluded that none of the available theories gave a satisfactory prediction of COD; a theory due to Chell[60] did, however, give correct values of plastic zone size under restricted conditions.

A fracture criterion based on the J-integral has attracted considerable attention as an alternative to COD testing. The method is based on the work of Rice on path-independent energy integrals in cracked structures[61, 62]. In elastic materials, J_I is simply the strain energy release rate, generalised to include non-linear stress-strain behaviour, so that in linear elastic materials J_{IC} and G_{IC} are identical. The use of the J-integral to define fracture criteria in ductile metals was developed by Begley and

Landes[63], and has since received wide attention. As yet, there are no published papers describing the direct measurement of J_{IC} in rubber-toughened plastics or other particulate composites. Williams and co-workers have attempted to estimate J_{IC} for HIPS and ABS indirectly, from the work done at the crack tip in SEN specimens subjected to tension or bending[57, 64]. As the fracture surfaces were extensively whitened, these estimates were probably high; the J_{IC} criterion should strictly be applied to the point of crack initiation rather than to later stages of ductile tearing.

V. Impact Tests

Both Izod and Charpy tests are in regular use for impact measurements on plastics. One disadvantage of the Izod test is that clamping forces introduce unknown stresses around the fracture zone, so that the test is less suitable than the Charpy test for research purposes. The Charpy test has been widely adopted for fracture mechanics measurements on both metals and polymers.

The Charpy test is essentially a high-speed three-point bending test. In a brittle material, the force exerted by the pendulum increases linearly with deflection until the strain energy release rate G_I reaches a critical value, and the crack begins to propagate. Once the crack has initiated, no further energy is required from the pendulum; crack propagation is maintained by energy already stored in the specimen. Thus the impact strength is basically a measure of the energy absorbed in bending the Charpy bar to the point of crack initiation; in addition, a small proportion of the energy abstracted from the pendulum is converted into kinetic energy of the two halves of the specimen. If the notch is sufficiently sharp, the test results are directly amenable to LEFM analysis. Standard test methods employ rounded notches, which are not entirely suitable. The more analytical studies that have been made recently, with the aim of obtaining fracture mechanics data for polymers, have used razor- or fatigue-notched specimens.

Under conditions of brittle fracture, the Charpy impact strength I is directly related to G_{IC} as follows[64]:

$$I - I_{ke} = BWC\left[\frac{dC}{d\left(\frac{a}{W}\right)}\right]^{-1} G_{IC} = BWZ\, G_{IC} \tag{15}$$

where B and W are the thickness and width of the specimen, a is crack length, and C is specimen compliance, defined as the ratio of deflection to applied load. The kinetic energy contribution to impact strength is given by I_{ke}. The equation may be tested by varying a, B, W, and the span between the specimen supports, and plotting the resulting values of I against BWZ. Experiments by Brown[65], Marshall et al.[66], and Plati and Williams[64] have shown that Eq. (15) successfully correlates Charpy impact data for sharply notched specimens of PS, PMMA, PVC, Nylon 66, polycarbonate, and all types of polyethylene over a wide range of specimen geometries. Of the materials tested in these programmes, only HIPS and ABS failed to give a linear relationship between I and BWZ. In the rubber-modified plastics, impact strength was proportional to ligament area, as follows:

$$I = B(W - a)R \qquad (16)$$

where R is the average energy absorbed per unit area of fracture surface.

The distinction between Eqs. (15) and (16) is an important one. Equation (16) describes a tough fracture, in which the elastic energy stored in the specimen at the point of crack initiation is insufficient to maintain crack propagation. If the kinetic energy of the pendulum is low, the crack may be brought to a stop. Usually, of course, the energy of the pendulum is in excess of that required to fracture the specimen, and additional energy is supplied to the crack as it extends. The energy required is roughly proportional to the fracture surface area; the kinetic energy term is smaller than in the brittle fracture case, and is negligible compared with the crack propagation energy.

Brown refers to the fracture surface energy R as a G_{IC} value[65]. This description is not entirely appropriate in view of the extensive plasticity at the crack tip, which is marked by stress-whitening. Plati and Williams suggest that R is equal to $J_{IC}/2$[64], but again the justification for equating the two quantities is weak. Further work will be required before the impact energy of HIPS and ABS can be related to clearly defined fracture mechanics parameters.

There is a transition in the impact behaviour of both HIPS and ABS at low temperatures. At 20 °C, the entire fracture surface is stress-whitened, indicating a tough fracture which cannot be analysed by LEFM methods. At lower temperatures, however, whitening is confined to a small area near the base of the notch, and experiments have shown that this type of fracture can be described by LEFM. The point is illustrated in Fig. 14, which presents data obtained by Bucknall and Reid[67] in Charpy tests on razor-notched HIPS specimens. At −25 °C, impact strength I is proportional to BWZ, as predicted by Eq. (15). At 40 °C, on the other hand, there

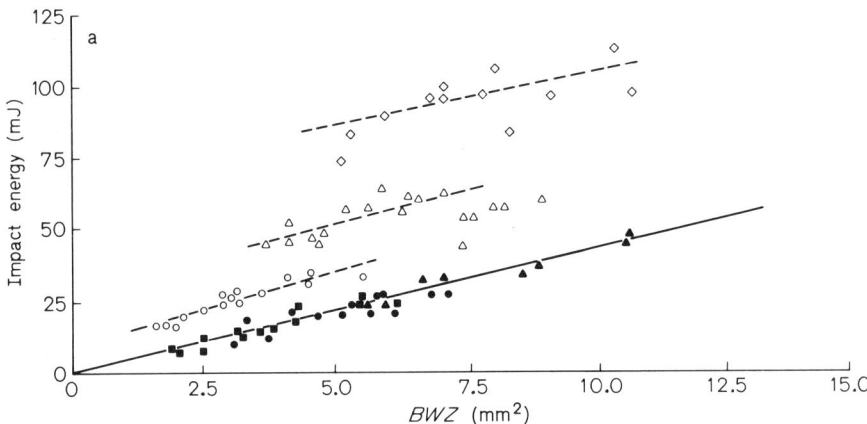

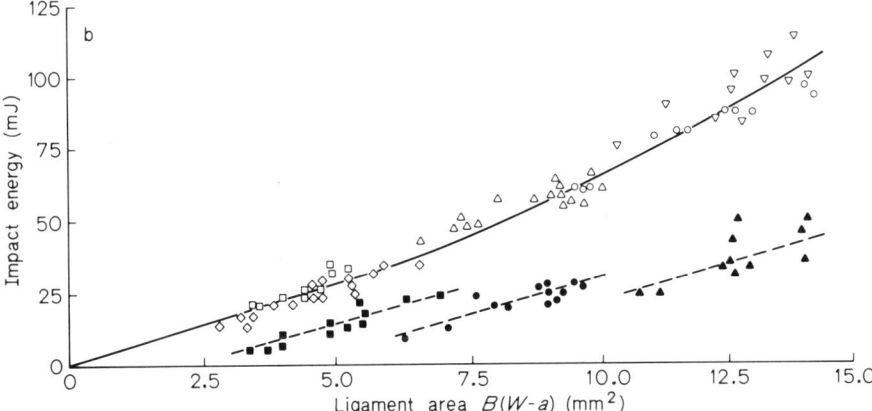

Fig. 14. Impact strengths of HIPS (a) plotted according to Eq. (15); and (b) plotted according to Eq. (16). Symbols show data for specimens of the same overall dimensions, but varying initial crack length: (◊, △, □, ▵, ○) at 40 °C; and (▲, ■, ●) at −25 °C

is no correlation between I and BWZ. Conversely, there is a correlation between I and fracture surface area $B(W - a)$ at 40 °C, as predicted by Eq. (16), but no correlation at −25 °C.

Figure 15 summarises the results of these experiments. Values of G_{IC} are given for temperatures up to 10 °C, above which LEFM methods do not apply, and values of R are given instead. Below the T_g of the rubber at about −90 °C, fracture is completely brittle, with no sign of stress-whitening, and a very low G_{IC}. Between −90° and 10°, stress-whitening occurs at the tip of the notch, and G_{IC} increases with temperature as the extent of the yield zone increases. It is clear from the work of Parvin and Williams[52] that fracture takes place under plane-stress rather than plane-strain conditions in the stress-whitened specimens. Away from the whitened zone,

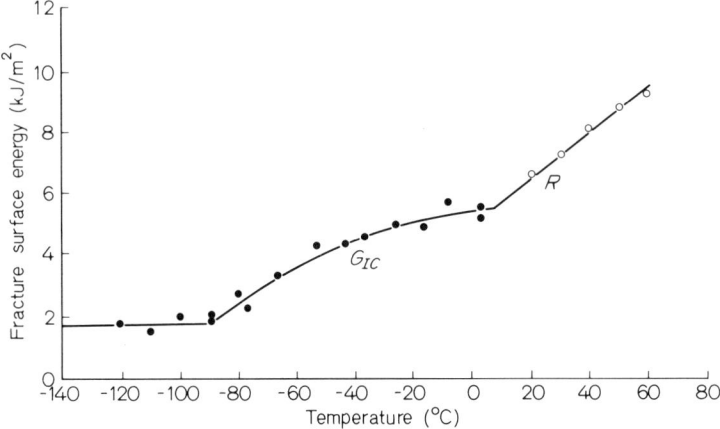

Fig. 15. Effects of temperature on fracture resistance of HIPS impact specimens

the appearance of the fracture surface is characteristic of a brittle, low-energy propagation, which can be related to the observed decrease in G_{IC} with crack speed in rubber-modified plastics. Above 10 °C, resistance to both initiation and propagation of a crack is high, and the entire fracture surface is whitened.

The increase in G_{IC} shown in Eq. (16) is due to a decrease in yield stress, which results in an increase in plastic zone size that is clearly visible as stress-whitening. The factors affecting yield stress in rubber-modified plastics are discussed in Section III. Temperature, strain rate, and rubber phase volume all control G_{IC} and I, through their influence upon yielding.

The foregoing fracture mechanics analysis of impact behaviour provides a useful basis for discussing weathering of rubber-modified plastics. The impact strengths of these materials tend to fall rapidly on exposure to ultraviolet light, including sunlight, owing to photo-oxidation of the rubber phase in the exposed surface layer of the material. Scott *et al.* have shown that the oxidation reaction causes the T_g of the rubber to rise as high as room temperature[68], so that the rubber content is effectively zero for a depth of up to 100 μm below the surface. The result is a sharp drop in G_{IC} in the surface layer. For temperatures at which fracture is initiation-controlled, the impact strength also drops, as predicted by Eq. (15). At higher temperatures, where the fracture is propagation-controlled, the oxidation of the surface layer has a much smaller effect, and the impact strength is little affected, in accordance with Eq. (16). A fuller discussion of aging is given in Ref.[3].

VI. Concluding Remarks

To date, much of the work on the fracture and failure of multiphase polymers and polymer composites has been largely of an empirical nature, with an emphasis on product evaluation. Relatively little attempt has been made to analyse the data systematically, or to compare one type of composite with another. Recent advances in fracture mechanics provide an improved basis for such a comparison, and some progress can now be made in correlating the data. As this review shows, there is ample scope for work in this area. The rewards for achieving a better understanding of the subject are not purely scientific: improved products will undoubtedly result from a clearer understanding of the relationship between structure and fracture toughness in particulate composites.

VII. References

[1] Manson, J. A., Sperling, L. H.: Polymer blends and composites. New York: Plenum Press 1976
[2] Crowson, R. J., Arridge, R. G. C.: J. Mater. Sci. *12*, 2154 (1977)
[3] Bucknall, C. B.: Toughened plastics. London: Applied Science 1977
[4] Wagner, E. R., Robeson, L. M.: Rubber Chem. Technol. *43*, 1129 (1970)

5) Speri, W. M., Patrick, G. R.: Poly. Engng. Sci. *15*, 668 (1975)
6) Pramuk, P. F.: Poly. Engng. Sci. *16*, 559 (1976).
7) Bowden, P. B., Raha, S.: Phil. Mag. *22*, 463 (1970)
8) Whitney, W.: J. Appl. Phys. *34*, 3633 (1963).
9) Argon, A. S., Andrews, R. D., Godrick, J. A., Whitney, W.: J. Appl. Phys. *39*, 1899 (1968)
10) Bowden, P. B., in: The physics of glassy polymers. Haward, R. N. (ed.). London: Applied Science 1973
11) Ward, I. M.: Mechanical properties of solid polymers. New York: Wiley 1971
12) Sternstein, S. S., Ongchin, L.: Am. Chem. Soc. Polym. Prepr. *10* (2), 1117 (1969)
13) Kambour, R. P.: J. Poly. Sci. *D7*, 1 (1973)
14) Rabinowitz, S., Beardmore, P.: CRC Crit. Revs. Macromol. Sci. *1*, 1 (1972)
15) Oxborough, R. J., Bowden, P. B.: Phil. Mag. *28*, 547 (1973)
16) Bucknall, C. B., Smith, R. R.: Polymer *6*, 437 (1965)
17) Bucknall, C. B., Clayton, D.: J. Mater. Sci. *7*, 202 (1972)
18) Bucknall, C. B., Clayton, D., Keast, W. E.: J. Mater. Sci. *7*, 1443 (1972)
19) Bucknall, C. B., Clayton, D., Keast, W. E.: J. Mater. Sci. *8*, 514 (1973)
20) Bucknall, C. B., Drinkwater, I. C.: J. Mater. Sci. *8*, 1800 (1973)
21) Bucknall, C. B., Page, C. J., Young, V. O.: Am. Chem. Soc. Adv. Chem. Ser. *154*, 179 (1976)
22) Bucknall, C. B., Yoshii, T.: Br. Poly. J.: *10*, 53 (1978)
23) Bucknall, C. B., in: Polymer blends. Paul. D. R., Newman, S. (eds.). New York: Academic Press, in the press
24) Beardmore, P., Rabinowitz, S.: J. Mater. Sci. *10*, 1763 (1975)
25) Fenelon, P. J., Wilson, J. R.: Am. Chem. Soc. Adv. Chem. Ser. *154*, 247 (1976)
26) Cessna, L. C.: Poly. Engng. Sci. *14*, 696 (1974)
27) Nicolais, L.: Poly. Engng. Sci. *15*, 137 (1975)
28) Nicolais, L., Narkis, M.: Poly. Engng. Sci. *11*, 194 (1971)
29) Lavengood, R. E., Nicolais, L., Narkis, M.: J. Appl. Poly. Sci. *17*, 1173 (1973)
30) Eyring, H.: J. Chem. Phys. *4*, 283 (1936)
31) Bauwens-Crowet, C., Bauwens, J., Homes, G.: J. Poly. Sci. *A27*, 735 (1969)
32) Goodier, J. N.: J. Appl. Mech. *55*, 39 (1933)
33) Ishai, O., Cohen, L. J.: J. Compos. Mat. *2*, 302 (1968)
34) Nicolais, L., Narkis, M.: Poly. Engng. Sci. *13*, 469 (1973)
35) Nicolais, L., Acierno, D., Janacek, J.: Poly. Engng. Sci. *15*, 35 (1975)
36) Nicolais, L., DiBenedetto, A. T.: Int. J. Polym. Mater. *2*, 251 (1973)
37) Morbitzer, L., Kranz, D., Humme, G., Ott, K. H.: J. Appl. Poly. Sci. *20*, 2691 (1976)
38) Oxborough, R. J., Bowden, P. B.: Phil. Mag. *30*, 171 (1974)
39) Roylance, M. E., Roylance, D. K., Sultan, J. N.: Am. Chem. Soc. Adv. Chem. Ser. *154*, 192 (1976)
40) Imasawa, Y., Matsuo, M.: Poly. Engng. Sci. *10*, 261 (1970)
41) Matsuo, M., Nozaki, C., Jyo, Y.: Poly. Engng. Sci. *9*, 197 (1969)
42) Young, R. J., Beaumont, P. W. R.: J. Mater. Sci. *12*, 684 (1977)
43) Fowlkes, C. W.: Int. J. Fract. *10*, 99 (1974)
44) Anderton, G. E.: J. Appl. Poly. Sci. *19*, 3355 (1975)
45) Akhurst, S. R., Bucknall, C. B.: unpublished observations
46) Hobbs, S. Y.: J. Appl. Phys. *48*, 4052 (1977)
47) Wambach, A., Trachte, K., DiBenedetto, A.: J. Compos. Mat. *2*, 266 (1968)
48) Trachte, K., DiBenedetto, A.: Int. J. Polym. Mater. *1*, 75 (1971)
49) DiBenedetto, A., Wambach, A. D.: Int. J. Polym. Mater. *1*, 159 (1971)
50) DiBenedetto, A.: J. Macromol. Sci. Phys. *B7*, 657 (1973)
51) Lange, F. F., Radford, K. C.: J. Mater. Sci. *6*, 1197 (1971)
52) Parvin, M., Williams, J. G.: J. Mater. Sci. *11*, 2045 (1976)
53) Parvin, M., Williams, J. G.: J. Mater. Sci. *10*, 1883 (1975)
54) Kobayashi, T., Broutman, L. J.: J. Appl. Poly. Sci. *17*, 2053 (1973)
55) Bascom, W. D., Jones, R. L., Timmons, C. O., in: Adhesion science and technology, Vol. *9B*, Plenum Press, New York, 1976

[56] Burdekin, F. M., Stone, D. E. W.: J. Strain Anal. *1*, 145 (1966)
[57] Ferguson, R. J., Marshall, G. P., Williams, J. G.: Polymer *14*, 451 (1973)
[58] Dugdale, D. S.: J. Mech. Phys. Solids *8*, 100 (1960)
[59] Hoffman, R. D., Richmond, O.: J. Appl. Phys. *47*, 4289 (1976)
[60] Chell, G. G.: Mat. Sci. Engng. *17*, 227 (1975)
[61] Rice, J. R.: J. Appl. Mech. *35*, 379 (1968)
[62] Rice, J. R., in: Fracture II. Liebowitz, H. (ed.). New York: Academic Press 1968
[63] Begley, J. A., Landes, J. D.: ASTM STP *514*, 1, 24 (1972)
[64] Plati, E., Williams, J. G.: Poly. Engng. Sci. *15*, 470 (1975)
[65] Brown, H. R.: J. Mater. Sci. *8*, 941 (1973)
[66] Marshall, G. P., Williams, J. G., Turner, C. E.: J. Mater. Sci. *8*, 949 (1973)
[67] Bucknall, C. B., Reid, J. C.: unpublished observations
[68] Ghaffar, A., Scott, A., Scott, G.: Eur. Poly. J. *11*, 271 (1975)

Received December 30, 1977
J. D. Ferry (editor)

Author Index Volumes 1–27

Allegra, G. and *Bassi, I. W.:* Isomorphism in Synthetic Macromolecular Systems. Vol. 6, pp. 549–574.
Andrews, E. H.: Molecular Fracture in Polymers. Vol. 27, pp. 1–66.
Ayrey, G.: The Use of Isotopes in Polymer Analysis. Vol. 6, pp. 128–148.
Baldwin, R. L.: Sedimentation of High Polymers. Vol. 1, pp. 451–511.
Basedow, A. M. and *Ebert, K.:* Ultrasonic Degradation of Polymers in Solution. Vol. 22, pp. 83–148.
Batz, H.-G.: Polymeric Drugs. Vol. 23, pp. 25–53.
Bergsma, F. and *Kruissink, Ch. A.:* Ion-Exchange Membranes. Vol. 2, pp. 307–362.
Berry, G. C. and *Fox, T. G.:* The Viscosity of Polymers and Their Concentrated Solutions. Vol. 5, pp. 261–357.
Bevington, J. C.: Isotopic Methods in Polymer Chemistry. Vol. 2, pp. 1–17.
Bird, R. B., Warner, Jr., H. R., and *Evans, D. C.:* Kinetic Theory and Rheology of Dumbbell Suspensions with Brownian Motion. Vol. 8, pp. 1–90.
Böhm, L. L., Chmelir̆, M., Löhr, G., Schmitt, B. J. und *Schulz, G. V.:* Zustände und Reaktionen des Carbanions bei der anionischen Polymerisation des Styrols. Vol. 9, pp. 1–45.
Bovey, F. A. and *Tiers, G. V. D.:* The High Resolution Nuclear Magnetic Resonance Spectroscopy of Polymers. Vol. 3, pp. 139–195.
Braun, J.-M. and *Guillet, J. E.:* Study of Polymers by Inverse Gas Chromatography. Vol. 21, pp. 107–145.
Breitenbach, J. W., Olaj, O. F. und *Sommer, F.:* Polymerisationsanregung durch Elektrolyse. Vol. 9, pp. 47–227.
Bresler, S. E. and *Kazbekov, E. N.:* Macroradical Reactivity Studied by Electron Spin Resonance. Vol. 3, pp. 688–711.
Bucknall, C. B.: Fracture and Failure of Multiphase Polymers and Polymer Composites. Vol. 27, pp. 121–148.
Bywater, S.: Polymerization Initiated by Lithium and Its Compounds. Vol. 4, pp. 66–110.
Carrick, W. L.: The Mechanism of Olefin Polymerization by Ziegler-Natta Catalysts. Vol. 12, pp. 65–86.
Casale, A. and *Porter, R. S.:* Mechanical Synthesis of Block and Graft Copolymers. Vol. 17, pp. 1–71
Cerf, R.: La dynamique des solutions de macromolecules dans un champ de vitesses. Vol. 1, pp. 382–450.
Cicchetti, O.: Mechanisms of Oxidative Photodegradation and of UV Stabilization of Polyolefins. Vol. 7, pp. 70–112.
Clark, D. T.: ESCA Applied to Polymers. Vol. 24, pp. 125–188.
Coleman, Jr., L. E. and *Meinhardt, N. A.:* Polymerization Reactions of Vinyl Ketones. Vol. 1, pp. 159–179.
Crescenzi, V.: Some Recent Studies of Polyelectrolyte Solutions. Vol. 5, pp. 358–386.
Davydov, B. E. and *Krentsel, B. A.:* Progress in the Chemistry of Polyconjugated Systems. Vol. 25, pp. 1–46.
Dole, M.: Calorimetric Studies of States and Transitions in Solid High Polymers. Vol. 2, pp. 221–274.

Dreyfuss, P. and *Dreyfuss, M. P.:* Polytetrahydrofuran. Vol. 4, pp. 528–590.
Dušek, K. and *Prins, W.:* Structure and Elasticity of Non-Crystalline Polymer Networks. Vol. 6, pp. 1–102.
Eastham, A. M.: Some Aspects of the Polymerization of Cyclic Ethers. Vol. 2, pp. 18–50.
Ehrlich, P. and *Mortimer, G. A.:* Fundamentals of the Free-Radical Polymerization of Ethylene. Vol. 7, pp. 386–448.
Eisenberg, A.: Ionic Forces in Polymers. Vol. 5, pp. 59–112.
Elias, H.-G., Bareiss, R. und *Watterson, J. G.:* Mittelwerte des Molekulargewichts und anderer Eigenschaften. Vol. 11, pp. 111–204.
Fischer, H.: Freie Radikale während der Polymerisation, nachgewiesen und identifiziert durch Elektronenspinresonanz. Vol. 5, pp. 463–530.
Fujita, H.: Diffusion in Polymer-Diluent Systems. Vol. 3, pp. 1–47.
Funke, W.: Über die Strukturaufklärung vernetzter Makromoleküle, insbesondere vernetzter Polyesterharze, mit chemischen Methoden. Vol. 4, pp. 157–235.
Gal'braikh, L. S. and *Rogovin, Z. A.:* Chemical Transformations of Cellulose. Vol. 14, pp. 87–130.
Gandini, A.: The Behaviour of Furan Derivatives in Polymerization Reactions. Vol. 25, pp. 47–96.
Gerrens, H.: Kinetik der Emulsionspolymerisation. Vol. 1, pp. 234–328.
Goethals, E. J.: The Formation of Cyclic Oligomers in the Cationic Polymerization of Heterocycles. Vol. 23, pp. 103–130.
Graessley, W. W.: The Entanglement Concept in Polymer Rheology. Vol. 16, pp. 1–179.
Hay, A. S.: Aromatic Polyethers. Vol. 4, pp. 496–527.
Hayakawa, R. and *Wada, Y.:* Piezoelectricity and Related Properties of Polymer Films. Vol. 11, pp. 1–55.
Heitz, W.: Polymeric Reagents. Polymer Design, Scope, and Limitations. Vol. 23, pp. 1–23.
Helfferich, F.: Ionenaustausch. Vol. 1, pp. 329–381.
Hendra, P. J.: Laser-Raman Spectra of Polymers. Vol. 6, pp. 151–169.
Henrici-Olivé, G. und *Olivé, S.:* Kettenübertragung bei der radikalischen Polymerisation. Vol. 2, pp. 496–577.
Henrici-Olivé, G. und *Olivé, S.:* Koordinative Polymerisation an löslichen Übergangsmetall-Katalysatoren. Vol. 6, pp. 421–472.
Henrici-Olivé, G. and *Olivé, S.:* Oligomerization of Ethylene with Soluble Transition-Metal Catalysts. Vol. 15, pp. 1–30.
Hermans, Jr., J., Lohr, D., and *Ferro, D.:* Treatment of the Folding and Unfolding of Protein Molecules in Solution According to a Lattic Model. Vol. 9, pp. 229–283.
Holzmüller, W.: Molecular Mobility, Deformation and Relaxation Processes in Polymers. Vol. 26, pp. 1–62.
Hutchison, J. and *Ledwith, A.:* Photoinitiation of Vinyl Polymerization by Aromatic Carbonyl Compounds. Vol. 14, pp. 49–86.
Iizuka, E.: Properties of Liquid Crystals of Polypeptides : with Stress on the Electromagnetic Orientation. Vol. 20, pp. 79–107.
Imanishi, Y.: Syntheses, Conformation, and Reactions of Cyclic Peptides. Vol. 20, pp. 1–77.
Inagaki, H.: Polymer Separation and Characterization by Thin-Layer Chromatography. Vol. 24, pp. 189–237.
Inoue, S.: Asymmetric Reactions of Synthetic Polypeptides. Vol. 21, pp. 77–106.
Ise, N.: Polymerizations under an Electric Field. Vol. 6, pp. 347–376.
Ise, N.: The Mean Activity Coefficient of Polyelectrolytes in Aqueous Solutions and Its Related Properties. Vol. 7, pp. 536–593.
Isihara, A.: Intramolecular Statistics of a Flexible Chain Molecule. Vol. 7, pp. 449–476.
Isihara, A.: Irreversible Processes in Solutions of Chain Polymers. Vol. 5, pp. 531–567.
Isihara, A. and *Guth, E.:* Theory of Dilute Macromolecular Solutions. Vol. 5, pp. 233–260.
Janeschitz-Kriegl, H.: Flow Birefringence of Elastico-Viscous Polymer Systems. Vol. 6, pp. 170–318.
Jenngins, B. R.: Electro-Optic Methods for Characterizing Macromolecules in Dilute Solution. Vol. 22, pp. 61–81.

Kawabata, S. and *Kawai, H.:* Strain Energy Density Functions of Rubber Vulcanizates from Biaxial Extension. Vol. 24, pp. 89–124.
Kennedy, J. P. and *Chou, T.:* Poly(isobutylene-*co*-β-Pinene): A New Sulfur Vulcanizable, Ozone Resistant Elastomer by Cationic Isomerization Copolymerization. Vol. 21, pp. 1–39.
Kennedy, J. P. and *Gillham, J. K.:* Cationic Polymerization of Olefins with Alkylaluminium Initators. Vol. 10, pp. 1–33.
Kennedy, J. P. and *Johnston, J. E.:* The Cationic Isomerization Polymerization of 3-Methyl-1-butene and 4-Methyl-1-pentene. Vol. 19, pp. 57–95.
Kennedy, J. P. and *Langer, Jr., A. W.:* Recent Advances in Cationic Polymerization. Vol. 3, pp. 508–580.
Kennedy, J. P. and *Otsu, T.:* Polymerization with Isomerization of Monomer Preceding Propagation. Vol. 7, pp. 369–385.
Kennedy, J. P. and *Rengachary, S.:* Correlation Between Cationic Model and Polymerization Reactions of Olefins. Vol. 14, pp. 1–48.
Kissin, Yu. V.: Structures of Copolymers of High Olefins. Vol. 15, pp. 91–155.
Kitagawa, T. and *Miyazawa, T.:* Neutron Scattering and Normal Vibrations of Polymers. Vol. 9, pp. 335–414.
Kitamaru, R. and *Horii, F.:* NMR Approach to the Phase Structure of Linear Polyethylene. Vol. 26., pp. 139–180.
Knappe, W.: Wärmeleitung in Polymeren. Vol. 7, pp. 477–535.
Koningsveld, R.: Preparative and Analytical Aspects of Polymer Fractionation. Vol. 7,
Kovacs, A. J.: Transition vitreuse dans les polymers amorphes. Etude phénoménologique. Vol. 3, pp. 394–507.
Krässig, H. A.: Graft Co-Polymerization of Cellulose and Its Derivatives. Vol. 4, pp. 111–156.
Kraus, G.: Reinforcement of Elastomers by Carbon Black. Vol. 8, pp. 155–237.
Krimm, S.: Infrared Spectra of High Polymers. Vol. 2, pp. 51–72.
Kuhn, W., Ramel, A., Walters, D. H., Ebner, G. and *Kuhn, H. J.:* The Production of Mechanical Energy from Different Forms of Chemical Energy with Homogeneous and Cross-Striated High Polymer Systems. Vol. 1, pp. 540–592.
Kunitake, T. and *Okahata, Y.:* Catalytic Hydrolysis by Synthetic Polymers. Vol. 20, pp.159–221.
Kurata, M. and *Stockmayer, W. H.:* Intrinsic Viscosities and Unperturbed Dimensions of Long Chain Molecules. Vol. 3, pp. 196–312.
Ledwith, A. and *Sherrington, D. C.:* Stable Organic Cation Salts: Ion Pair Equilibria and Use in Cationic Polymerization. Vol. 19, pp. 1–56.
Lee, C.-D. S. and *Daly, W. H.:* Mercaptan-Containing Polymers. Vol. 15, pp. 61–90.
Lipatov, Y. S.: Relaxation and Viscoelastic Properties of Heterogeneous Polymeric Compositions. Vol. 22, pp. 1–59.
Lipatov, Y. S.: The Iso-Free-Volume State and Glass Transitions in Amorphous Polymers: New Development of the Theory. Vol. 26, pp. 63–105.
Mano, E. B. and *Coutinho, F. M. B.:* Grafting on Polyamides. Vol. 19, pp. 97–116.
Meyerhoff, G.: Die viscosimetrische Molekulargewichtsbestimmung von Polymeren. Vol. 3, pp. 59–105.
Millich, F.: Rigid Rods and the Characterization of Polyisocyanides. Vol. 19, pp. 117–141.
Morawetz, H.: Specific Ion Binding by Polyelectrolytes. Vol. 1, pp. 1–34.
Mulvaney, J. E., Oversberger, C. C., and *Schiller, A. M.:* Anionic Polymerization. Vol. 3, pp. 106–138.
Okubo, T. and *Ise, N.:* Synthetic Polyelectrolytes as Models of Nucleic Acids and Esterases. Vol. 25, pp. 135–181.
Osaki, K.: Viscoelastic Properties of Dilute Polymer Solutions. Vol. 12, pp. 1–64.
Oster, G. and *Nishijima, Y.:* Fluorescence Methods in Polymer Science. Vol. 3, pp. 313–331.
Overberger, C. G. and *Moore, J. A.:* Ladder Polymers. Vol. 7, pp. 113–150.
Patat, F., Killmann, E. und *Schiebener, C.:* Die Absorption von Makromolekülen aus Lösung. Vol. 3, pp. 332–393.
Peticolas, W.L.: Inelastic Laser Light Scattering from Biological and Synthetic Polymers. Vol. 9, pp. 285–333.

Pino, P.: Optically Active Addition Polymers. Vol. 4, pp. 393–456.

Plesch, P. H.: The Propagation Rate-Constants in Cationic Polymerisations. Vol. 8, pp. 137–154.

Porod, G.: Anwendung und Ergebnisse der Röntgenkleinwinkelstreuung in festen Hochpolymeren. Vol. 2, pp. 363–400.

Postelnek, W., Coleman, L. E., and *Lovelace, A. M.:* Fluorine-Containing Polymers. I. Fluorinated Vinyl Polymers with Functional Groups, Condensation Polymers, and Styrene Polymers. Vol. 1, pp. 75–113.

Rempp, P., Herz, J., and *Borchard, W.:* Model Networks. Vol. 26, pp. 107–137.

Rogovin, Z. A. and *Gabrielyan, G. A.:* Chemical Modifications of Fibre Forming Polymers and Copolymers of Acrylonitrile. Vol. 25, pp. 97–134.

Roha, M.: Ionic Factors in Steric Control. Vol. 4, pp. 353–392.

Roha, M.: The Chemistry of Coordinate Polymerization of Dienes. Vol. 1, pp. 512–539.

Safford, G. J. and *Naumann, A. W.:* Low Frequency Motions in Polymers as Measured by Neutron Inelastic Scattering. Vol. 5, pp. 1–27.

Schuerch, C.: The Chemical Synthesis and Properties of Polysaccharides of Biomedical Interest. Vol. 10, pp. 173–194.

Schulz, R. C. und *Kaiser, E.:* Synthese und Eigenschaften von optisch aktiven Polymeren. Vol. 4, pp. 236–315.

Seanor, D. A.: Charge Transfer in Polymers. Vol. 4, pp. 317–352.

Seidl, J., Malinský, J., Dušek, K. und *Heitz, W.:* Makroporöse Styrol-Divinylbenzol-Copolymere und ihre Verwendung in der Chromatographie und zur Darstellung von Ionenaustauschern. Vol. 5, pp. 113–213.

Semjonow, V.: Schmelzviskositäten hochpolymerer Stoffe. Vol. 5, pp. 387–450.

Semlyen, J. A.: Ring-Chain Equilibria and the Conformations of Polymer Chains. Vol. 21, pp. 41–75.

Sharkey, W. H.: Polymerizations Through the Carbon-Sulphur Double Bond. Vol. 17, pp. 73–103.

Shimidzu, T.: Cooperative Actions in the Nucleophile-Containing Polymers. Vol. 23, pp. 55–102.

Slichter, W. P.: The Study of High Polymers by Nuclear Magnetic Resonance. Vol. 1, pp. 35–74.

Small, P. A.: Long-Chain Branching in Polymers. Vol. 18, pp. 1–64.

Smets, G.: Block and Graft Copolymers. Vol. 2, pp. 173–220.

Sohma, J. and *Sakaguchi, M.:* ESR Studies on Polymer Radicals Produced by Mechanical Destruction and Their Reactivity. Vol. 20, pp. 109–158.

Sotobayashi, H. und *Springer, J.:* Oligomere in verdünnten Lösungen. Vol. 6, pp. 473–548.

Sperati, C. A. and *Starkweather, Jr., H. W.:* Fluorine-Containing Polymers. II. Polytetrafluoroethylene. Vol. 2, pp. 465–495.

Sprung, M. M.: Recent Progress in Silicone Chemistry. I. Hydrolysis of Reactive Silane Intermediates. Vol. 2, pp. 442–464.

Stille, J. K.: Diels-Alder Polymerization. Vol. 3, pp. 48–58.

Szwarc, M.: Termination of Anionic Polymerization. Vol. 2, pp. 275–306.

Szwarc, M.: The Kinetics and Mechanism of N-carboxy-α-amino-acid Anhydride (NCA) Polymerization to Poly-amino Acids. Vol. 4, pp. 1–65.

Szwarc, M.: Thermodynamics of Polymerization with Special Emphasis on Living Polymers. Vol. 4, pp. 457–495.

Tani, H.: Stereospecific Polymerization of Aldehydes and Epoxides. Vol. 11, pp. 57–110.

Tate, B. E.: Polymerization of Itaconic Acid and Derivatives. Vol. 5, pp. 214–232.

Tazuke, S.: Photosensitized Charge Transfer Polymerization. Vol. 6, pp. 321–346.

Teramoto, A. and *Fujita, H.:* Conformation-dependent Properties of Synthetic Polypeptides in the Helix-Coil Transition Region. Vol. 18, pp. 65–149.

Thomas, W. M.: Mechanism of Acrylonitrile Polymerization. Vol. 2, pp. 401–441.

Tobolsky, A. V. and *DuPré, D. B.:* Macromolecular Relaxation in the Damped Torsional Oscillator and Statistical Segment Models. Vol. 6, pp. 103–127.

Tosi, C. and *Ciampelli, F.:* Applications of Infrared Spectroscopy to Ethylene-Propylene Copolymers. Vol. 12, pp. 87–130.

Tosi, C.: Sequence Distribution in Copolymers: Numerical Tables. Vol. 5, pp. 451–462.

Tsuchida, E. and *Nishide, H.:* Polymer-Metal Complexes and Their Catalytic Activity. Vol. 24, pp. 1–87.
Tsuji, K.: ESR Study of Photodegradation of Polymers. Vol. 12, pp. 131–190.
Valvassori, A. and *Sartori, G.:* Present Status of the Multicomponent Copolymerization Theory. Vol. 5, pp. 28–58.
Voorn, M. J.: Phase Separation in Polymer Solutions. Vol. 1, pp. 192–233.
Werber, F. X.: Polymerization of Olefins on Supported Catalysts. Vol. 1, pp. 180–191.
Wichterle, O., Šebenda, J., and *Králíček, J.:* The Anionic Polymerization of Caprolactam. Vol. 2, pp. 578–595.
Wilkes, G. L.: The Measurement of Molecular Orientation in Polymeric Solids. Vol. 8, pp. 91–136.
Williams, J. G.: Applications of Linear Fracture Mechanics. Vol. 27, pp. 67–120.
Wöhrle, D.: Polymere aus Nitrilen. Vol. 10, pp. 35–107.
Wolf, B. A.: Zur Thermodynamik der enthalpisch und der entropisch bedingten Entmischung von Polymerlösungen. Vol. 10, pp. 109–171.
Woodward, A. E. and *Sauer, J. A.:* The Dynamic Mechanical Properties of High Polymers at Low Temperatures. Vol. 1, pp. 114–158.
Wunderlich, B. and *Baur, H.:* Heat Capacities of Linear High Polymers. Vol. 7, pp. 151–368.
Wunderlich, B.: Crystallization During Polymerization. Vol. 5, pp. 568–619.
Wrasidlo, W.: Thermal Analysis of Polymers. Vol. 13, pp. 1–99.
Yamazaki, N.: Electrolytically Initiated Polymerization. Vol. 6, pp. 377–400.
Yoshida, H. and *Hayashi, K.:* Initiation Process of Radiation-induced Ionic Polymerization as Studied by Electron Spin Resonance. Vol. 6, pp. 401–420.
Zachmann, H. G.: Das Kristallisations- und Schmelzverhalten hochpolymerer Stoffe. Vol. 3, pp. 581–687.
Zambelli, A. and *Tosi, C.:* Stereochemistry of Propylene Polymerization. Vol. 15, pp. 31–60.

Polymers

Properties and Applications

Editorial Board:
H.-J. Cantow,
H.J. Harwood,
J.P. Kennedy, J. Meißner,
S. Okamura, G. Olivé,
S. Olivé

Springer-Verlag
Berlin
Heidelberg
New York

Volume 1
B. Rånby, J.F. Rabek

ESR Spectroscopy in Polymer Research

1977. 356 figures, 29 tables. XIV, 410 pages
ISBN 3-540-08151-8

The main purpose of this book is to collect the present available information on the applications of electron spin resonance (ESR) spectroscopy in polymer research. The book has been written both for those who want an introduction to this field, and for those who are already familiar with ESR and are interested in application to polymers. Therefore, the fundamental principles of ESR spectroscopy are first outlined, the experimental methods including computer applications are described in more detail, and the main emphasis is on the application of ESR methods to polymer problems. The authors hope that this book will provide a useful source of information by giving a coherent treatment and extensive references to original papers, reviews, and discussions in monographs and books. In this way we hope to encourage polymer chemists, organic chemists, biochemists, physicists, and material scientists to apply ESR methods to their research problems. (2519 references).

Volume 2
H.-H. Kausch

Polymer Fracture

1978. Approx. 350 pages
ISBN 3-540-08786-9

In the last fifteen years modern spectroscopical methods (ESR, IR) and conventional methods of structure research have permitted considerable progress in the investigation of deformation and fracture of polymeric materials. For the first time in western languages a unified view of the kinetic theory of polymer fracture is presented by one of the scientists contributing to its development.

Advances in Polymer Science

Fortschritte der Hochpolymeren-Forschung

Editors: H.-J. Cantow, G. Dall'Asta, J. D. Ferry, H. Fujita, M. Gordon, W. Kern, G. Natta, S. Okamura, C. G. Overberger, T. Saegusa, G. V. Schulz, W. P. Slichter, A. J. Staverman, J. K. Stille

Volume 21
Mechanisms of Polyreactions – Polymer Characterization
1976. 68 figures. III, 161 pages
ISBN 3-540-07727-8
Contents:
J. P. Kennedy, T. Chou: Poly (Isobutylene-co-ß-Pinene): A New Sulfur Vulcanizable, Ozone Resistant Elastomer by Cationic Isomerization Copolymerization
J. A. Semlyen: Ring-Chain Equilibria and the Conformations of Polymer Chains
S. Inoue: Asymmetric Reactions of Synthetic Polypeptides
J.-M. Braun, J. E. Guillet: Study of Polymers by Inverse Gas Chromatography

Volume 22
Physical Chemistry
1977. 77 figures. III, 153 pages
ISBN 3-540-07942-4
Contents:
Y. S. Lipatov: Relaxation and Vixcolelastic Properties of Heterogeneous Polymeric Compositions
B. R. Jennings: Electro-Optic Methods for Characterizing Macromolecules in Dilute Solution
A. M. Basedow, K. Ebert: Ultrasonic Degradation of Polymers in Solution

Volume 23
Reactivities
1977. 28 figures, 29 tables. III, 136 pages
ISBN 3-540-07943-2
Contents:
T. Shimidzu: Cooperative Actions in the Nucleophile-Containing Polymers
H.-G. Batz: Polymeric Drugs
E. J. Goethals: The Formation of Cyclic Oligomers in the Cationic: Polymerization of Heterocycles
W. Heitz: Polymeric Reagents. Polymer Design, Scope and Limitations

Volume 24
Molecular Properties
1977. 130 figures, 33 tables. III, 244 pages
ISBN 3-540-08124-0
Contents:
E. Tsuchida, H. Nishide: Polymer Metal Complexes and Their Catalytic Activity
S. Kawabata, H. Kawai: Strain Energy Density Functions of Rubber Vulcanizates from Biaxial Extension
D. T. Clark: ESCA Applied to Polymers
H. Inagaki: Polymer Separation and Characterization by Thin-Layer Chromatography

Volume 25
Polymer Chemistry
1977. 55 figures. VII, 187 pages
ISBN 3-540-08389-8
Contents:
B. E. Davydov, B. A. Krentsel: Progress in the Chemistry of Polyconjugated Systems
A. Gandini: The Behaviour of Furan Derivatives in Polymerization Reactions
Z. A. Rogovin, G. A. Gabrielyan: Chemical Modifications of Fibre Forming Polymers and Copolymers of Acrylonitrile
T. Okubo, N. Ise: Synthetic Polyelectrolytes as Models of Nucleic Acids and Esterases

Volume 26
Conformation and Morphology
1978. 61 figures. IV, 185 pages
ISBN 3-540-08677-3
Contents:
W. Holzmüller: Molecular Mobility, Deformation and Relaxation Processes in Polymers
Y. Lipatov: The Iso-Free-Volume State and Glass Transitions in Amorphous Polymers: New Development of the Theory
J. E. Herz, P. Rempp, W. Borchard: Model Networks
R. Kitamaru, F. Horii: NMR Approach to the Phase Structure of Linear Polyethylene

Springer-Verlag
Berlin Heidelberg New York